A Guide for the Implementation and Use of Chloramines

Awwa
Research
Foundation

About the Awwa Research Foundation

The Awwa Research Foundation (AwwaRF) is a member-supported, international, nonprofit organization that sponsors research to enable water utilities, public health agencies, and other professionals to provide safe and affordable drinking water to consumers.

The Foundation's mission is to advance the science of water to improve the quality of life. To achieve this mission, the Foundation sponsors studies on all aspects of drinking water, including supply and resources, treatment, monitoring and analysis, distribution, management, and health effects. Funding for research is provided primarily by subscription payments from approximately 1,000 utilities, consulting firms, and manufacturers in North America and abroad. Additional funding comes from collaborative partnerships with other national and international organizations, allowing for resources to be leveraged, expertise to be shared, and broad-based knowledge to be developed and disseminated. Government funding serves as a third source of research dollars.

From its headquarters in Denver, Colorado, the Foundation's staff directs and supports the efforts of more than 800 volunteers who serve on the board of trustees and various committees. These volunteers represent many facets of the water industry, and contribute their expertise to select and monitor research studies that benefit the entire drinking water community.

The results of research are disseminated through a number of channels, including reports, the Web site, conferences, and periodicals.

For subscribers, the Foundation serves as a cooperative program in which water suppliers unite to pool their resources. By applying Foundation research findings, these water suppliers can save substantial costs and stay on the leading edge of drinking water science and technology. Since its inception, AwwaRF has supplied the water community with more than $300 million in applied research.

More information about the Foundation and how to become a subscriber is available on the Web at **www.awwarf.org**.

A Guide for the Implementation and Use of Chloramines

Prepared by:
Leland L. Harms
Black & Veatch Corporation, Kansas City, MO 64114
and
Christine Owen
Tampa Bay Water, Clearwater, FL 33761-3970

Jointly sponsored by:
Awwa Research Foundation
6666 West Quincy Avenue, Denver, CO 80235-3098
and
Tampa Bay Water
Clearwater, FL 33761-3970

Published by:

Awwa Research Foundation
Advancing the Science of Water®

DISCLAIMER

This study was jointly funded by the Awwa Research Foundation (AwwaRF) and Tampa Bay Water. AwwaRF and Tampa Bay Water assume no responsibility for the content of the research study reported in this publication or for the opinions or statements of fact expressed in the report. The mention of trade names for commercial products does not represent or imply the approval or endorsement of AwwaRF or Tampa Bay Water. This report is presented solely for informational purposes.

Printed on recycled paper

CONTENTS

TABLES

FIGURES

FOREWORD

The Awwa Research Foundation is a nonprofit corporation that is dedicated to the implementation of a research effort to help utilities respond to regulatory requirements and traditional high-priority concerns of the industry. The research agenda is developed through a process of consultation with subscribers and drinking water professionals. Under the umbrella of a Strategic Research plan, the Research Advisory Council prioritizes the suggested projects based upon current and future needs, applicability, and past work; the recommendations are forwarded to the Board of Trustees for final selection. The foundation also sponsors research projects through the unsolicited proposal process; the Collaborative Research, Research Applications, and Tailored Collaboration programs; and various joint research efforts with organizations such as the U.S. Environmental Protection Agency, the U.S. Bureau of Reclamation, and the Association of California Water Agencies.

This publication is a result of one of these sponsored studies, and it is hoped that its findings will be applied in communities throughout the world. The following report serves not only as a means of communicating the results of the water industry's centralized research program but also as a tool to enlist the further support of the nonmember utilities and individuals.

Projects are managed closely from their inception to the final report by the foundation's staff and large cadre of volunteers who willingly contribute their time and expertise. The foundation serves a planning and management function and awards contracts to other institutions such as water utilities, universities, and engineering firms. The funding for this research effort comes primarily from the Subscription Program, through which water utilities subscribe to the research program and make an annual payment proportionate to the volume of water they deliver and consultants and manufacturers subscribe based on their annual billings. The program offers a cost-effective and fair method for funding research in the public interest.

A broad spectrum of water supply issues is addressed by the foundation's research agenda: resources, treatment and operations, distribution and storage, water quality and analysis, toxicology, economics, and management. The ultimate purpose of the coordinated effort is to assist water suppliers to provide the highest possible quality of water economically and reliably. The true benefits are realized when the results are implemented at the utility level. The foundation's trustees are pleased to offer this publication as a contribution toward that end.

The objectives of this study were to research advances in chloramination theory and practice and to summarize this information in a comprehensive guidance manual. It is hoped that the report will serve as a guide to utilities currently using or considering the use of chloramination in their treatment process.

Walter J. Bishop
Chair, Board of Trustees
Awwa Research Foundation

James F. Manwaring, P.E.
Executive Director
Awwa Research Foundation

ACKNOWLEDGMENTS

This study was made possible by financial support from the Awwa Research Foundation and Tampa Bay Water. The authors would like to express their appreciation to the Foundation, to Tampa Bay Water, and to our AwwaRF project officer, India Williams. The authors of this report are indebted to the many operators, managers, engineers, and plant superintendents who generously gave of their time and expertise. The time spent filling out the survey form, responding to questions on the telephone, and sharing experiences during the site visits was appreciated. We say thank you to each of the 63 utilities that participated in this project.

EXECUTIVE SUMMARY

BACKGROUND

Chloramines have been used for disinfection for years in some water systems. Most recently they have been relegated primarily as a secondary disinfectant for surface water systems because of disinfection requirements promulgated under the Surface Water Treatment Rule. As limits on disinfection byproducts become even more stringent, many more water systems will be evaluating and/or implementing chloramines in their treatment process.

A primary purpose of this project was to develop material that could be used as a concise and practical guidance manual for those responsible for chloramination at water works facilities. To this end, this report briefly discusses the most important concepts using language that hopefully is easy to understand and apply by both engineering personnel and operating staff. Additional material and detail have been added as appendices for the more interested readers.

Chloramines are formed by the addition of chlorine and ammonia. At near neutral pH, chloramines are not as effective as free chlorine for inactivating pathogenic organisms. In today's practice, disinfection is most often carried out with free chlorine, chlorine dioxide, ozone, or UV with chloramines providing a protective residual within the distribution system. A primary advantage to chloramines is the reduced levels of disinfection byproducts (TTHM and HAA5) obtainable when compared to using only free chlorine throughout the entire system. Other advantages to chloramination include more persistent residuals, a reduction in taste and odor complaints, and biofilm control.

IMPORTANT CONCEPTS

Several important concepts regarding forming and using chloramines are discussed in the manual. Concerns regarding its biological efficiency and stimulus as a biological nutrient are put into perspective under normal operating conditions where a sufficient chloramine residual is maintained. Present information suggests that microbial growth is not greatly stimulated as ammonia releases during chloramine decay if a sufficient residual is maintained throughout the distribution system. In fact, several utilities indicated that using chloramines has assisted in controlling biofilms within the distribution system.

As with any oxidant, chloramines have disinfection byproducts; but the byproducts are not the chlorinated species currently regulated. The promulgation of SDWA regulations that lower allowable levels of DBPs will continue to cause more public water systems to implement chloramination.

A chief concern of many utilities using chloramines is nitrification. Nitrification is the process of ammonia being converted to nitrite and eventually nitrate. Problems caused when nitrification becomes excessive include low chlorine residuals, a drop in pH and/or alkalinity, and increased biological growth within distribution system piping. Problems with chlorine residual depletion and nitrification may be directly related to water age. Utilities that minimize the time water is stored prior to use appear to minimize disinfectant residual loss and nitrification.

In an effort to control nitrification, some water systems introduce free chlorine instead of chloramines within the distribution system periodically. During periods when free chlorine is used, DBP levels increase with the higher levels becoming a potential health or regulatory issue.

Blending of chlorinated water with chloraminated water can be a problem. Improper blending may produce tastes and odors objectionable to the consumer. Chloramines have also been implicated in the degradation of some materials used in the distribution system, but most utilities do not indicate this as a major problem.

Finally, technical personnel tend to neglect public perception regarding changing disinfectants. In addition to notifying and explaining chloramines satisfactorily, water department personnel must be aware that chloramines can interfere with some medical treatment (kidney dialysis) and metabolism of fish.

CHEMISTRY

Chloramines are formed when ammonia is mixed with chlorinated water. Although the chemistry of chloramination can be quite complex, inorganic chloramination can be explained simply as the formation of three compounds: monochloramine, dichloramine, and trichloramine or nitrogen trichloride. In practice, only monochloramine is desired and this compound is best formed by maintaining 3 to 5 mg/L of chlorine for every 1 mg/L of ammonia.

PRACTICE OF CHLORAMINATION

A critical element of successfully using chloramines lies in maintaining the proper chlorine to ammonia ratio. As the chlorine:ammonia-nitrogen ratio exceeds 5:1 because of too much chlorine, taste and odor causing compounds tend to form thus generating complaints from consumers. If too much ammonia is fed, free ammonia will enter the distribution system and increased biological growth and/or nitrification may result.

Ammonia is most often added as either aqua ammonia or anhydrous ammonia. Some operating difficulties have been noted with the chemical feed systems. Among these difficulties are diffusers becoming plugged or caked with material, storage and feeding issues when using pressurized tanks, and injecting into high pressure lines. None of these problems are insurmountable and chloramination becomes another routine operation at most facilities.

Producing high quality drinking water is the primary goal of every water treatment facility. The most serious water quality issue appears to be one of potential nitrification, as nitrification can reduce chlorine residuals to a non-safe level as well as promoting other water quality problems. Two primary groups of organisms are responsible for nitrification; they are generally grouped as ammonia-oxidizing bacteria, AOB, and nitrite-oxidizing bacteria, NOB. Because excess ammonia can stimulate biological growth, it is important to maintain an adequate chloramine residual in all parts of the system.

A distinct advantage of chloramines over free chlorine is the lower chlorinated DBPs a utility can achieve. Pure chloramines form negligible amounts of THMs and HAA5, both of which are now regulated under the Stage 1 DBPR. A potential problem for some utilities that switch to free chlorine periodically is that the Stage 2 DBPR may seriously limit the periodic use of free chlorine by requiring that DBP concentrations resulting from free chlorine be included when compliance is determined. The intent of the Stage 2 regulation is to include periods of highest DBP concentrations when compliance with the MCLs is determined so the pending rule will likely address this issue.

Although chloramines are known to be less effective for inactivation of pathogens, as the pH increases the difference in effectiveness between free chlorine and chloramines becomes less.

At a finished water pH of 7.5 or higher, the disinfection efficiency of chloramines begins to approach that of free chlorine and the difference is negligible by pH 9.

Of the 63 water systems surveyed, only six blended chloraminated and chlorinated water together. Of these six utilities, a few received some taste and odor complaints but most systems avoid blending by keeping chlorinated water separate from chloraminated water by feeding into different pressure zones. Metropolitan Water District of Southern California developed a computer model to assist in determining what fraction of blending could occur without causing consumer complaints.

Communication and coordination among parties affected by implementing or using chloramines are vital components of a successful operation. Multiple avenues of communication are recommended. These include informing regulators, making presentations to the general public, newsletters, media releases, pamphlets and providing a source where questions can be directed. All internal staff should be briefed on the important issues so they can be in agreement with the decision to use chloramines and be effective ambassadors to customers.

A critical population that must be made aware of chloramine conversion are dialysis centers. Chloramines interfere with dialysis but most medical centers routinely monitor the water used so this should not be a problem for them. Pet owners, primarily those with aquariums, should also be made aware of any changes in chlorination practice. Cards can be provided at pet stores to facilitate dissemination of this information.

Customers will have questions; be prepared to answer them. Even routine household activities will generate questions regarding chloramines. Activities such as pool maintenance, cleaning surfaces, icemakers, watering plants, processing foods and beverages, and doing laundry may lead someone to seek additional information regarding chloramination.

SURVEY RESULTS

Sixty-three water systems responded to the survey. Site visits were made to 13 locations with the locations being selected based on survey responses. The remaining utilities were contacted as necessary by telephone to clarify responses on the survey form or to gather additional information. The survey results reflect the state of chloramination practice in the U.S. and report the experience, opinions, and concerns of the water treatment professionals who practice chloramination daily.

The most common reason for using chloramines was to reduce DBP levels, and an average decrease of 82 percent in THMs was documented. Experience with chloramines ranged from one year to 75 years but most of the systems have employed chloramines for more than 15 years. Twenty-two percent of these utilities switch to free chlorine periodically in an effort to reduce biofilms and control nitrification. If free chlorine is used, it normally is for a fairly short duration of from one to four weeks, but one utility uses free chlorine for up to three months.

During the day to day practice of chloramination, very few operating problems were reported. The largest problem area reported was one of storing, mixing, or feeding either the chlorine or ammonia and about 11 percent of the utilities reported some problems maintaining their target chlorine:ammonia-nitrogen ratio. According to the systems surveyed, the objectives associated with chloramination are being achieved and the use of chloramines is a practical method of operating a treatment system.

Several appendices are included at the rear of the report to assist the individual who desires more detail and additional information. Appendices are:

- Appendix A, which contains a comprehensive review of the literature and a list of references should further reading be desired.
- Appendix B with a copy of the survey form sent to utilities during this study.
- Appendix C, a listing of water systems that participated in this study.
- Appendix D which provides a more detailed view of the chemistry of chloramines.
- Appendix E giving information obtained during the 13 site visits, and
- Appendix F which provides examples of materials used to communicate the conversion to chloramines.

CHAPTER 1
INTRODUCTION

Chloramination of drinking water is not a new technology; some water utilities have successfully used chloramines for over 40 years. Other utilities, however, have experienced problems including nitrification, the stimulation of biological growth within the distribution system, an inability to maintain adequate chlorine residuals throughout the entire system, and customer complaints about the taste and/or odor of the water. Today, probably more than ever before, many utilities are considering chloramines as a secondary disinfectant within their distribution system, primarily as an economical method to meet ever more stringent disinfection byproduct limits. This project was undertaken to assist long-term users of chloramines as well as those just beginning to implement this disinfection process.

OBJECTIVES

The objectives of this study were to identify new knowledge and information regarding the use of chloramines and to synthesize this information into step by step procedures for using and implementing chloramines in a water system. A primary objective of the work was to develop a user friendly guidance manual that would be suitable for every day use by treatment plant operators and other personnel responsible for the successful use of chloramines. Practical problems such as the need for free chlorine burns to mitigate nitrification, how to optimize and control the chlorine to ammonia ratio, how to meet new disinfection byproduct (DBP) limits while achieving satisfactory disinfection, and the blending of chlorinated water with chloraminated water were all issues to be investigated.

BACKGROUND

The use of chloramines entails the addition of ammonia to solutions with chlorine to eliminate hypochlorous acid, HOCl, and hypochlorite ion, OCl^-. Chloramines have been used since the early 1900s for disinfection, taste and odor control, and for maintaining chlorine residuals within the distribution system (Wolfe, Ward and Olson 1984). Chloramines are especially favored by utilities with large distribution systems as they tend to persist longer than free chlorine residuals. It has long been known that combined chlorine is not as active a biocide as free chlorine (Clarke et al. 1962), and the use of chloramines as a primary disinfectant has been largely discontinued since 1989 when the Surface Water Treatment Rule (SWTR) (USEPA 1989) was promulgated with minimum inactivation requirements for enteric viruses. Very long contact times are normally required for chloramines to inactivate viruses and most water treatment facilities do not have adequate detention time to allow the use of chloramines for this purpose. Chloramines continue to be widely used as a secondary disinfectant within the distribution system and some utilities are also able to satisfy primary disinfection with only chloramines.

Chloramines continue to gain popularity as water utilities seek treatment technologies that enable disinfectant byproduct regulations to be met while maintaining disinfection and providing suitable disinfectant residuals. Of the 63 water purveyors using chloramines that responded during this investigation, ten have used chloramines for 40 years or more but a similar number have only recently implemented chloramination.

Several issues regarding the use of chloramines became apparent as the use of combined chlorine was explored with the many public water systems responding to the survey developed for this project. The major issues and problems were:

1. Concerns regarding nitrification within the distribution system
2. Loss of chlorine residual at remote areas of the distribution system
3. Proper control of both ammonia and chlorine
4. Maintenance of chemical feed equipment
5. Degradation of some materials from reacting with chloramines
6. Meeting disinfection requirements while also keeping disinfection byproducts at a low level
7. Maintaining water quality within the distribution system
8. Blending water disinfected with free chlorine with water disinfected with chloramines
9. Public perception regarding the use of chloramines
10. How to notify the public when switching to chloramines or from chloramines back to free chlorine

This report attempts to answer the ten questions listed above. The first part of this report is intentionally kept brief in the hope that this portion will find its way onto the operating floor for use as a quick and easy, user friendly manual for operators charged with forming and applying chloramines. Supplemental material from the survey, results and trends, a literature review, and more details on chloramine chemistry can all be found in the latter sections of the report and the appendices.

CHAPTER 2
IMPORTANT CONCEPTS

A complete literature review is included as Appendix A to this report. This section discusses important concepts regarding chloramines found in the literature and encountered in practice. The interested reader can consult the references listed for each concept for additional information.

FORMATION

Regardless of whether chlorine is added as solid calcium hypochlorite, $Ca(OCl)_2$, liquid sodium chlorite, NaOCl, or chlorine gas, Cl_2, free chlorine in the form of hypochlorous acid, HOCl, and hypochlorite ion, OCl^-, are formed. Simplified chemical reactions are shown below:

$$Ca(OCl)_2 + 2H_2O \rightarrow Ca(OH)_2 + 2HOCl$$

$$NaOCl + H_2O \rightarrow NaOH + HOCl$$

$$Cl_2 + H_2O \rightarrow HOCl + HCl$$

$$HOCl \leftrightarrow H^+ + OCl^- \quad \text{and}$$

$$HCl \leftrightarrow H^+ + Cl^-$$

Note that when chlorine gas is used, the pH tends to drop due to the formation of hydrochloric acid, HCl; and when either sodium or calcium hypochlorite is used, the pH tends to rise due to the formation of hydroxides as end products of the reaction. Chlorine is a strong oxidizing agent and attacks organics in solution including bacteria and viruses. Chlorine gas, hypochlorous acid, and hypochlorite ion are collectively referred to as free chlorine.

Ammonium ion, NH_4^+, exists in solution in equilibrium with ammonia, NH_3, and the hydrogen ion. When ammonia is added to chlorinated water, the ammonia reacts with the chlorine to form chloramines. Chloramines are also referred to as combined chlorine. Chloramines may exist as monochloramine, NH_2Cl, dichloramine, $NHCl_2$, or trichloramine, NCl_3 (Faust and Aly 1998, Sawyer and McCarty 1978). For drinking water systems, chloramines are desired to be present in the monochloramine form as small amounts of dichloramine and trichloramine result in taste and odor complaints from consumers. A more detailed discussion of how to maintain monochloramine as the chloramine of choice is included in the next section on chloramine chemistry.

DISINFECTION

Free available chlorine is a more active disinfectant than chloramine (White 1986), and consequently chloramines are often used as secondary disinfectants within the distribution system. Most surface water systems and those ground water systems required to inactivate viruses use free chlorine or another disinfectant prior to chloramination. For example, if 0.5 log of *Giardia* and 2 log of virus inactivation are desired for a filtered water at 20°C and pH 7.5; this requirement would be satisfied by using 1 mg/L of free chlorine with an effective contact time of 11 min but the same dose would require 320 min for chloramine (USEPA 1991).

Chloramines are commonly used in large distribution systems as a secondary disinfectant. Although chloramines do decay with time (see Appendix D), at pH values of 7.5 or greater they are relatively stable and provide a consistent residual throughout the distribution system. Free ammonia, NH_3, and nitrogen gas are byproducts of chloramine decay, and because nitrogen is a micro-nutrient known to stimulate microbiological growth, there has been a concern that using chloramines may promote bacterial growth within the distribution system. This concern becomes even more important for those systems that also use a phosphate inhibitor to control corrosion or incrustation on filter media as phosphorus is the other primary micro-nutrient that stimulates biological growth. While most water systems using free chlorine maintain residuals on the order of 0.5 to 1.5 mg/L, chloramine residuals are normally higher. It is not unusual for a treatment plant using chloramines to pump to the distribution system with a 2 to 3.5 mg/L residual. Current evidence exists suggesting that chloramine decay does not stimulate microbial growth to unacceptable levels in the presence of a sufficient residual. One large water system using chloramines routinely has heterotrophic plate counts (HPC) counts of less than 10 cfu/mL when the chloramine residual is maintained at 2 mg/L or more.

The Stage 1 Disinfectants/Disinfection Byproducts Rule (Stage 1 DBPR) (USEPA 1998) limits concentrations of disinfectants, including chloramines, by establishing a Maximum Residual Disinfectant Level (MRDL) of 4 mg/L for both chlorine and chloramines. Compliance with the MRDL is based on a system average so some treatment facilities discharge to the system with a chloramine concentration of slightly greater than 4 mg/L.

DISINFECTION BYPRODUCTS

One reason for using chloramines is to reduce levels of chlorinated byproducts in the treated water. Regulated byproducts such as trihalomethanes, THMs, continue to form within the distribution system after leaving the treatment plant as long as free chlorine is present. This additional formation can largely be eliminated by using chloramines within the distribution system instead of free chlorine. The promulgation of the Stage 1 DBPR and the pending Stage 2 DBPR has prompted many utilities to consider chloramines as a treatment technology in order to limit their chlorinated byproducts and comply with MCLs for THMs and haloacetic acids (HAA5).

Although chloramines do not form substantial levels of halogenated byproducts, they have byproducts of their own. Any oxidation-reduction reaction will have byproducts from that reaction; consequently, any chemical oxidant used for disinfection will produce some byproducts. Currently byproducts from chloramination are not regulated either because they are not considered to be health hazards or because present information is inadequate to make a risk assessment. Byproducts of chloramination for which some concern has been expressed include cyanogen chloride and N-nitrosodimethylamine (NDMA).

NITRIFICATION

Nitrification is the process of conversion of ammonia-nitrogen to nitrite and nitrate. This oxidation process is primarily conducted by two groups of microorganisms called ammonia-oxidizing bacteria (AOB) and nitrite-oxidizing bacteria (NOB). AOB organisms use ammonia as an energy source in their metabolic processes and in so doing convert the ammonia to nitrite. If nitrites were allowed to accumulate unhindered, they would themselves be toxic to the AOB and the reaction would cease. NOB organisms, however, use nitrite for their energy with nitrate being the final end product of their metabolism. If other environmental conditions are suitable, these organisms can cause nitrification to proceed within the distribution system. Undesirable results of nitrification can be the build-up of nitrate and nitrite; loss of chloramine residual; water age; decreases in pH, alkalinity, and dissolved oxygen, and an increase in bacterial growth.

Some water systems practice a "free chlorine burn" periodically as a means of combating nitrification. In this practice, chloramination is discontinued from a few days for up to a month; and during this time, free chlorine is fed into the distribution system. Using a free chlorine burn reportedly restores chlorine residuals to acceptable levels. Disadvantages to this practice include customer complaints because they discern a difference in how their water tastes, and the increase in disinfection byproducts such as THMs and HAA5. Usually DBP monitoring is not conducted within the distribution system while using free chlorine as DBP levels would be expected to be much higher than normal.

Nitrification is a serious problem for some public water systems. One medium size water system needed to go into the distribution system with over 7 mg/L of chloramines in order to maintain a satisfactory residual at the ends of their distribution system. Operating in this fashion is not allowable under the Stage 1 DBPR which sets a MRDL of 4 mg/L for chloramines.

Water systems that use a free chlorine burn as a means of nitrification control, may have difficulty meeting Stage 2 DBP criteria as one requirement is to sample during the month that highest concentrations of DBPs are expected. Some provisions for not sampling during this particular month will be included in the Rule, and careful application of free chlorine between scheduled sampling events should be allowed. Presently, it is not known if State Primacy Agents will routinely excuse monitoring during a free chlorine burn.

BLENDING

Many water systems use a combination of ground water and treated surface water as the community water supply. Ground water is often supplied from multiple wells which are treated only with chlorine for disinfection and discharge directly into the distribution system. If chloramines are used for secondary disinfection from the surface water treatment facility, chlorinated ground water may then be blended within the distribution system with chloraminated water. Depending upon the blended percentage, this practice can lead to taste and odor problems as the combined residual may tend to form as dichloramine instead of monochloramine. The reader is referred to the breakpoint chlorination curve in the section on Chloramine Chemistry as an aid to understanding this point. Small concentrations of dichloramine are noticeable to many consumers and frequent complaints of taste and odor may result.

As noted later in this report, some systems are able to blend chloraminated water with chlorinated water successfully without substantial consumer complaints. Other systems have gone to isolating different zones within the distribution systems so the two different waters stay separate for the most part. Another option is to implement chloramination at each well site so only chloraminated water is sent into the distribution system from both wells and the treatment plant.

ELASTOMER DEGRADATION

Information in the literature indicates that chloramines may be responsible for accelerated degradation of some materials (Reiber 1993). Evidence seems to exist that implicates chloramines in the rapid deterioration of gaskets, compression fittings, valve components, and other items within a distribution system exposed to combined chlorine. It is difficult to document this issue in actual practice, but it is enough of a problem that it was noticed by about 20 of the utilities contacted. Of the 63 water systems participating in this study, four identified this degradation as a problem and eight said it was one of their concerns.

PUBLIC NOTICE AND ACCEPTANCE

Conversion to chloramines must not be done in a vacuum. The public must be kept fully aware of any changes and given an indication of what to expect. Certainly, public health officials and medical centers must be notified as chloramines can adversely affect some medical procedures such as kidney dialysis. Also, pet shops and home aquarium users need to be notified in as much as possible. Chloramine residuals and ammonia can be toxic to fish and other aquatic life at low concentrations and provision for using chloraminated water must be made by the individual pet owner. Also, since chloramines do not dissipate as quickly as free chlorine, they can kill fish directly.

Often any change in treating water is perceived by the public in a negative way. In recent years, the public has become more conscious and knowledgeable of the quality of its drinking water and more concerned about the security and safety of its supply. Public acceptance is not automatic and any public water supply switching to chloramines needs to develop a comprehensive public awareness and information program.

CHAPTER 3
CHEMISTRY OF CHLORAMINES

This short section on the chemistry of chloramines is intended to provide a brief introduction to a complex subject. Only the formation of inorganic chloramines will be discussed here. For more detailed information, the reader is referred to Appendix D of this report, the references listed in the section on Formation in Chapter 2, Wolfe, Ward, and Olson (1984), and White (1986).

Ammonia, NH_3, is mixed with chlorinated water to form chloramines. Ammonia is normally added to drinking water as aqua ammonia, NH_4OH, or as anhydrous ammonia, a liquified gas. It is also sometimes obtained in solid form as either ammonium sulfate or ammonium chloride but this approach is generally used only for smaller systems. Ammonia is in equilibrium with the ammonium ion and this ionized species is the predominant form in most drinking water systems:

$$NH_3 + H_2O \leftrightarrow NH_4^+ + OH^-$$

The relationship between free ammonia, NH_3, and ammonium ion, NH_4, is highly pH dependent. For water systems with a pH between 7 and 8.5, the predominant form is ammonium ion; as the pH increases, a higher percentage of ammonia gas is formed until at pH 11 over 95% will be present as NH_3.

Chloramines exist in water as three compounds:

NH_2Cl	monochloramine
$NHCl_2$	dichloramine, and
NCl_3	trichloramine/nitrogen trichloride

The two primary forms of interest are mono- and dichloramine. Under normal operating conditions, trichloramine will not be present. The preferred residual is monochloramine as even trace amounts of dichloramine often give objectionable tastes and odors to the water.

Monochloramine forms as chlorine reacts with the ammonia. Figure 3.1 illustrates the relationship between chlorine dose, chlorine residual, and the form of chloramines for a hypothetical water containing 0.5 mg/L NH_3-N. Normally sufficient ammonia is added to the disinfected water to produce a chlorine to ammonia ratio of between 3:1 and 5:1. This ratio is expressed as mg/L Cl_2: mg/L NH_3-N and is based on the chlorine residual, not the chlorine dose.

The chlorine demand of the water is defined as the difference between the chlorine dose and the chlorine residual. In the example shown in Figure 3.1, a chlorine dose of about 2.7 mg/L is necessary to produce a desired residual of 2.2 mg/L because for this set of conditions the chlorine demand is about 0.5 mg/L. The chlorine demand varies with contact time, temperature, pH, and chlorine dose. Thus, to properly operate with chloramines, the operator will need knowledge regarding the chlorine demand for the water being treated for the present condition. For a particular solution, the chloramine residual will reach its maximum concentration at approximately a 5:1 chlorine:ammonia ratio. Monochloramine is formed at a chlorine:ammonia ratio of 5:1 or less, and increasing the chlorine:ammonia ratio greater than 5:1 will cause favorable conditions for the undesirable formation of dichloramine. Note that if the chlorine:ammonia ratio is increased beyond 5:1, the total chlorine residual will actually decrease even though the chlorine dose is increasing. This is an important point for operators who are attempting to maintain a chloramine residual within a specified range. If an operator notices that the chloramine residual is less than desired, the normal first reaction is to turn up the feed rate on the chlorinator to increase the chlorine dose. However, this response may lead to a further drop in the total chlorine residual (see Figure 3.1), and the operator should first calculate the current chlorine:ammonia ratio to ensure which portion of the chlorine curve is applicable before adjusting any chemical feed systems.

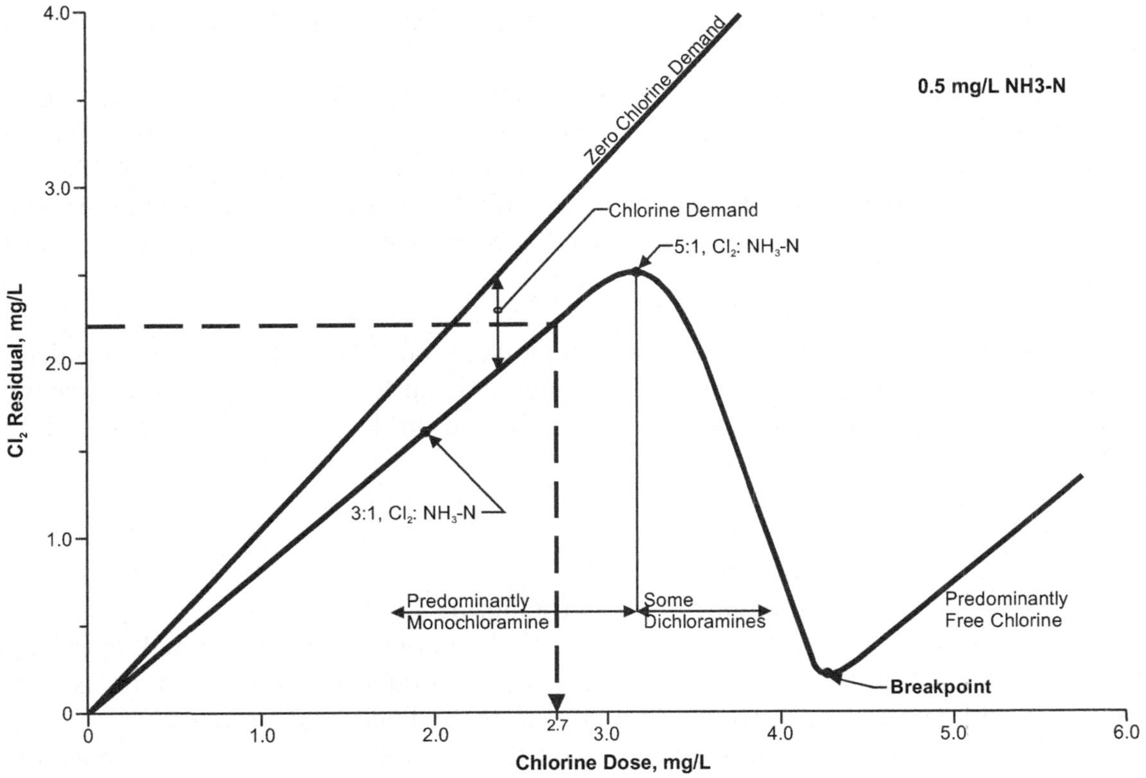

Figure 3.1 Breakpoint chlorination graph

CHAPTER 4
PRACTICE OF CHLORAMINATION

This chapter discusses issues surrounding the practical application of chloramines for a drinking water system. The text is intended as a guide from fundamental concepts to the successful application of chloramines. Additional references are provided within the text as sources for more in-depth discussions for the interested reader.

CHLORINE:AMMONIA RATIO

Controlling the chlorine and ammonia doses to provide the appropriate chlorine:ammonia ratio in the treated water is an important and basic component for the successful use of chloramines. As explained earlier, most water systems operate with a chlorine:ammonia ratio of 3:1 to 5:1 by weight. In other words, for every mg/L of ammonia in the water, 3 to 5 mg/L of chlorine should be present. Thus, it is the chlorine residual, not the chlorine dose, that is used to compute the actual chlorine:ammonia ratio.

The desired form of chloramines in drinking water, monochloramine, is formed when the chlorine:ammonia ratio is less than about 5:1 (see Figure 3.1). In practice, there is normally a desired target for the chlorine residual leaving the treatment plant and entering the distribution system with the intent of maintaining an adequate chloramine residual in all parts of the system. Each distribution system is different, but most utilities keep a chloramine residual of from 1.0 to 2.5 mg/L within the system. With the target residual set with the chlorine demand considered, the question then becomes one of how much ammonia to feed to convert the free chlorine residual to chloramine. Again referring to Figure 3.1, the actual chlorine demand must be determined by each water system, and the demand will vary from day to day depending upon changes in water quality as well as with other factors mentioned previously.

If insufficient ammonia is fed (Cl_2:NH_3-N > 5:1), the residual will not be fully converted to monochloramine and the formation of some dichloramines is likely. As mentioned previously, low concentrations of dichloramines are known to be objectionable to consumers and should be avoided. Additionally, if the Cl_2:NH_3-N exceeds 7.6:1, substantial amounts of free chlorine will be present which negates the decision for using chloramines in the first place, for example limiting the formation of chlorinated byproducts such as THMs and HAA5.

If excess ammonia is fed, full formation of monochloramine should be attained. However, excessive amounts of free ammonia are thought to contribute to nitrification within the distribution system as ammonia is the initial compound necessary to cause AOB and NOB growth. Most of the systems surveyed target free ammonia concentrations leaving the treatment plant from < 0.1 to 0.3 mg/L NH_3-N. Measuring the free ammonia concentration after implementing chloramines will assist the operator in adjusting the Cl_2:NH_3-N ratio and maintaining the residual as monochloramine. Normally once the chemical feed systems for both ammonia and chlorine have been operated together for a few weeks, the plant operating staff will be able to form monochloramine with low ammonia residuals and only occasional checks of the ammonia concentration will be necessary. Measuring ammonia using a low-level, hand-held colorimeter is ordinarily satisfactory to control the Cl_2:NH_3-N ratio. More precise measurements can be obtained using a laboratory spectrophotometer or a specific ion electrode. On-line instruments for the continuous measurement of ammonia are also available.

OPERATIONS

Ammonia can be added in several different forms; aqua ammonia and anhydrous ammonia are commonly used, but ammonium sulfate is used at several locations. As with any chemical feed system, some operating problems can occur. Almost all the operating difficulties reported dealt with the ammonia feed systems, not chlorine. The long and successful history of feeding chlorine within the water works industry is presumed to be the reason that most chlorine feed systems are considered to be trouble free and any necessary maintenance or repairs are regarded as routine.

Aqua ammonia is frequently stored outdoors with transfer pumps moving the solution to day tanks where metering pumps are used to add ammonia to the process stream. One utility with bulk storage tank, day tanks, and all transfer and metering pumps in one interior room notes that persistent small leaks cause noticeable fumes within the room. A separate exhaust system was installed in this location to ensure adequate ventilation.

Some feed systems use pressurized tanks to feed the ammonia to day tanks or metering pumps. Sunlight can affect the pressure within the tank. High concentrations of aqua ammonia will vaporize at high temperatures and may cause the feed pumps to vapor lock; consequently most systems use concentrations in the 20% range. One utility sprays water over their storage tanks during the summer months to keep the tanks cool. Conversely, one system experiences problems during extremely cold weather when the pressure drops too low and feed rates are restricted.

Small pipes under pressure and diffusers are both used to inject ammonia into the process stream. Deposits may form with either approach as ammonia is a basic chemical and can cause calcium precipitates to form in localized areas. One system uses ammoniators and the ammonia injector tubes need to be cleaned with dilute acid approximately every other month. When the ammonia feed line goes directly into a header under pressure, the correct shut down procedure needs to be followed to avoid water backing up into the transfer line.

The majority of systems contacted still use free chlorine to ensure disinfection and ammonia is added downstream after sufficient contact time with free chlorine has been achieved. Monitoring for both free and total chlorine residuals is recommended as an aid to detecting problems. If only total chlorine is measured, the need to adjust the ammonia feed system will not be noticed. Additionally, some locations are able to determine improper hydraulic splits between treatment trains using this technique.

Several systems noted that chloramines were responsible for the degradation of gasket material and other components within their system. Chloramine resistant plumbing materials are available for homeowners and plumbers at local hardware and plumbing supply stores. When specifically asked about operating problems using chloramines, no systems mentioned this issue.

WATER QUALITY ISSUES

Providing treated water that meets all state and federal regulations, is safe to drink, and is aesthetically pleasing are necessary goals for any water supplier. This section will discuss the water quality issues peculiar to using chloramines and will provide operating guidance for meeting all water quality goals.

Nitrification

The potential for nitrification within the distribution system is perhaps the single most prevalent water quality concern among those using chloramines or considering implementing chloramination. Understanding the importance of nitrification, what causes it, and what can prevent it are all important for the operator of a chloraminated water system to understand.

Some autotrophic microorganisms use ammonia-nitrogen as a food source to generate energy for their metabolic processes. End products from this oxidation-reduction reaction are nitrite and nitrate, and this microbial conversion of ammonia to nitrite and nitrate is called nitrification. This conversion is normally considered to occur in two stages with two somewhat different groups of organisms responsible for each stage. During the first stage, bacteria oxidize ammonia, NH_3, to nitrite, NO_2^-. The group of organisms capable of doing the first stage oxidation are loosely referred to as ammonia-oxidizing bacteria, AOB. In the second stage, the nitrite formed from the first stage oxidation is itself oxidized to nitrate, NO_3, by another group of organisms referred to as nitrite-oxidizing bacteria or NOB.

Adverse impacts from nitrification on finished water quality include the following:

- Diminished chlorine residuals. A loss of chloramine residual is often used as a primary indicator of nitrification. The authors know of a medium size water system that needed to feed more than 7 mg/L of chlorine in order to maintain an acceptable residual within their distribution system.
- Increased biological growth. Nitrogen and phosphorus are two primary micro-nutrients needed to stimulate biological growth. Excess nitrogen, especially in the form of nitrate, is readily available for most microorganisms to use in growth processes. The available organic carbon in a treated drinking water may also limit biological activity. The microbial quality of the treated water is normally routinely monitored using HPC counts. If nitrification is occurring, the death of nitrifying organisms may cause an increase in HPC levels as nitrifying organisms release the carbon stored in their biomass. Since the HPC levels of water systems vary from < 10/mL to > 200/mL, a prudent action is to regularly monitor for HPC to determine if an increase is occurring.
- Loss of dissolved oxygen. Normally the dissolved oxygen content of drinking water is not a water quality issue as the water is aerated when delivered at the faucet. However, in cases of extremely low DO, other chemical reactions can occur such as within the sulfur cycle. Obnoxious tastes and odors have been produced in some water supplies in areas of low DO, especially if coupled with low chlorine residuals.
- Increase in nitrite and nitrate. Both nitrite and nitrate have an MCL under the Safe Drinking Water Act regulations. These MCLs are 1.0 mg/L and 10 mg/L as N, respectively. These limits would not be exceeded in drinking water because of nitrification as compliance is determined by sampling the plant effluent not at the tap. However, because the public health is at issue here, it is wise to limit both of these contaminants to less than the MCL at all locations within the distribution system. Nitrite is actually toxic to most microorganisms and can be toxic to even the AOB group if excessive accumulation of nitrite were to occur. Nitrite is known to cause a rapid decrease in monochloramine levels (Wilczak et al. 1996).
- Reduction in pH and alkalinity. Both AOB and NOB biological reactions produce acid thus causing a reduction in pH and alkalinity. A drop of pH within a distribution system of a few tenths is not considered abnormal, but if pH and alkalinity begin to drop more

than usual, nitrification could be the cause. Maintaining the system pH and alkalinity within a normal operating range is very important to keeping a stable and non-corrosive water available for consumers.

The adverse effects on water quality itemized above are usually managed by maintaining sufficient chloramine residuals throughout all parts of the distribution system, limiting water age, and monitoring for nitrification. Because microbial growth is temperature dependent, systems that carry warm water are more subject to the effects of nitrification and more vigilance is necessary in warmer climates.

Some systems practice a free chlorine burn to discourage the growth of NOB and AOB. This free chlorine burn can last from a few days to a month or more but in itself, it can also create water quality issues. Switching to free chlorine increases the amount of chlorinated disinfection byproducts. As discussed later, this may not be allowed under the Stage 2 regulations. Systems that employ this practice are predominately located in warmer climates such as Florida and Texas.

Keys to maintaining a distribution system without nitrification problems are:

- Know your system. Monitor the plant discharge on a regular basis to establish baseline values for ammonia, nitrate, nitrite, HPC, pH, alkalinity, and chloramine residuals. Compare values obtained from the plant effluent with a set of stations within the distribution system. Additional information on monitoring programs can be found in the write-ups from site visits; see Appendix B.
- Maintain low levels of ammonia. A small amount of excess/free ammonia is to be expected in the plant effluent but limit this excess to < 0.3 mg/L as NH_3-N as much as possible. Some well operated facilities are able to control free ammonia in the plant discharge to about 0.1 mg/L NH_3-N.
- Increase monitoring during periods of warmer water. As mentioned elsewhere, warm water is more conducive to microbial activity with resulting nitrification issues.
- Maintain a sufficient chloramine residual in all parts of the distribution system. Most systems reporting problems with nitrification were found to have chloramine levels less than 2 mg/L (Wilczak et al. 1996).
- Limit water age. The primary responsibility for treatment plant operators is to ensure sufficient water is available for use. Consequently, the effects of storage and transportation on water quality are often ignored. Excessive water age can contribute to water quality problems including nitrification as more time is available for microbial activity. Although sufficient water must be maintained within the system for all uses including fire demands, there can be a tendency to keep all storage tanks full even when demands are low. Many utilities have developed models of their distribution system to assist with limiting water age as well as other water quality related problems.

Some practical experience on controlling nitrification is documented in the notes from the site visits (see Appendix E). Note especially the monitoring conducted by System I, and the problem that developed in Wichita Falls during a mild season when water age was higher than normal. As nitrites increased to 2 mg/L, a local restaurant began having difficulty with “pink chicken”. Apparently the preservatives used by their chicken supplier reacted with the nitrites to turn the chicken pink even though it was well cooked.

Disinfection

Most facilities do not use chloramines as a primary disinfectant because of the long contact times required for virus and *Giardia* inactivation. Disinfection is normally accomplished with free chlorine, chlorine dioxide, or ozone and chloramines are applied as a secondary disinfectant to avoid subsequent contamination within the distribution system. To avoid the chlorinated disinfection byproducts associated with free chlorine, some utilities are installing post-filter UV to satisfy disinfection requirements. Unfortunately, with the lack of current guidance from EPA regarding inactivation using UV, some states are allowing the units to be installed but are not yet granting any credit for inactivation. Of the 63 utilities surveyed, the most common method of achieving disinfection was the application of free chlorine prior to filtration with ammonia being added somewhere after the filters, usually after a clear well, contact basin, or reservoir. This scenario allows the required CT to be obtained using free chlorine and then chloramines are used to maintain a residual within the distribution system. For reference, Table 4.1 compares required CT values to achieve 0.5 log *Giardia* inactivation and 2 log virus inactivation for chlorine, chloramines, and chlorine dioxide for a sampling of water quality conditions.

Table 4.1
CT requirements for chlorine, chloramines, and chlorine dioxide

		CT Required, (mg/L)(min)		
pH units	Temperature °C	Chlorine $Cl_2 = 1$ mg/L	Chloramines	Chlorine Dioxide
7	5	25	857	5.6
7	10	19	643	4.2
7	20	9	321	2.5
8	5	36	857	5.6
8	10	27	643	4.2
8	20	14	321	2.5
9	5	52	857	5.6
9	10	39	321	4.2
9	20	20	643	2.5

Source: USEPA 1991

Maintaining adequate residuals within the system is important when operating with combined chlorine. Most utilities surveyed targeted 1 mg/L or more as the desired residual at the ends of the distribution system. Although relatively stable, some chloramine decay can be expected and the target residual into the distribution system is frequently more than 1.5 mg/L. The Surface Water Treatment Rule (SWTR) requires that a trace chlorine residual be maintained at all locations within the distribution system. Some states, however, have more stringent requirements when chloramines are used. Disinfection criteria for Kansas, for example, require a residual of 1.0 mg/L at the ends of the distribution system (KS Dept. of Health and Environment 1995).

Disinfection Byproduct Control

A primary reason that many utilities are switching to chloramines is that it becomes an economical option for controlling chlorinated byproducts. Pure chloramines in a stable water form only minimal amounts of chlorinated byproducts such as THMs and HAA5 (USEPA 1981). Experience has shown, however, that a change in pH may affect chlorinated DBP concentrations. However, when compared to DBP levels generated by free chlorine, chloramines substantially reduce these concentrations.

A pending problem related to DBP control for systems that use chloramines, is the option of conducting periodic free chlorine burns to mitigate nitrification. Several systems, predominately those with warmer water, use this technique on a regular basis. The Stage 2 DBPR, expected to become final near the end of 2004, may eliminate this option. An initial draft of the Stage 2 DBPR, based on the Stage 2 M-DBP Agreement in Principle developed jointly by the USEPA and its Federal Advisory Committee, indicates that each PWS must schedule DBP sampling during the peak historical month for TTHM and HAA5 unless rationale is provided for choosing another month. Because the intent of this provision is to include monitoring data during periods of high DBP concentrations with samples being collected at locations that include locations with the highest known concentrations of TTHM and HAA5, it seems unlikely that a PWS could avoid sampling during a period of free chlorine use by employing a rationale that such sampling would yield high DBP data and potentially cause a violation of the MCLs. Several state agencies were contacted regarding this issue. In general, state primacy agents have not considered this problem yet and will await publication of the final rule and any guidance that EPA will make available.

Any oxidation-reduction reaction will form some byproducts of that reaction so the presence of some disinfection byproducts from chloramination is to be expected. Researchers have reported trace amounts of many byproducts including chloropicrin, chloral hydrate, N-Nitrosodimethylamine (NDMA) and cyanogen chloride. Currently there are no MCLs for these byproducts and none are pending. The two most likely candidates for regulation are thought to be NDMA and cyanogen chloride but neither contaminant is included on EPA's Drinking Water Contaminant Candidate List (USEPA 1998). For cyanogen chloride, it is a reaction byproduct of free chlorine, ammonia, and organics present in the water. According to the World Health Organization, concentrations detected in drinking water are very low, on the order of 0.5 to 1.5 μg/L (WHO 1993). The public health concern regarding cyanogen chloride is much broader than its presence in drinking water. Cyanogen chloride is metabolozed to cyanide in humans and the health concern is based on total cyanide in the human body, not on just cyanogen chloride. WHO expressed a guideline value of 70 μg/L for cyanide as a safe limit.

NDMA has been traced to ion exchange water treatment processes and cationic polymers as well as being a byproduct of chloramination (CA Department of Health Services 2002). Of 64 samples collected from systems using chloramines, NDMA concentrations ranged from near zero to 63.7 ng/L (ppt) with a mean concentration of 1.8 ng/L. Twenty-three percent of the 64 samples had concentrations of NDMA below detection levels of 1 ppt. Nitrification is thought to contribute to NDMA levels so programs to control nitrification will also be beneficial for controlling NDMA. At this time, NDMA from chloramination is not projected to be regulated.

Bromide

Chloramination in the presence of bromide is not completely understood but it is well documented that several compounds may be produced from the reaction of chlorine, ammonia, and bromide. Both bromamines (NH_2Br, $NHBr_2$, and NBr_3) as well as bromochloramine (NHBrCl) have been found (Wajon and Morris 1980, Symons et al. 1998). Bromamines tend to form more rapidly than chloramines but also degrade more rapidly. The Metropolitan Water District of Southern California reported an increase in THMs from 10 μg/L to 20 μg/L while using chloramines during a drought when the bromide level in the raw water increased to 0.5 mg/L.

In addition to affecting DBP concentrations and speciation, water with bromide will also interfere with chlorine residual measurements. A common method of determining chlorine residuals utilizes DPD, *N,N*-diethyl-*p*-phenylenediamine. Valentine (1986) studied the use of DPD in the presence of bromochloramine. He concluded that bromochloramine reacts as half bromamine and half chloramine thereby being recorded in both free chlorine and total chlorine measurements.

Bacteriological Quality

It has been well documented that chloramines are not as active a disinfectant as free chlorine. This information has been incorporated into the disinfection requirements of the SDWA regulations and has limited chloramine use to one of a secondary disinfectant at most locations. However, chloramines do deter biological growth and are nearly as active as free chlorine in many water treatment systems.

Figure 4.1 (White 1986) shows the classic diagram often used as proof of the inferior germicidal properties of monochloramine when compared to free chlorine (hypochlorous acid and hypochlorite ion). The free chlorine fractions of hypochlorous acid and hypochlorite ion are considered to be in equilibrium with each other according to the equations given below:

$$Cl_2 + H_2O \leftrightarrow HOCl + H^+ + Cl^- \quad \text{and}$$

$$HOCl \leftrightarrow H^+ + OCl^-$$

Figure 4.2 illustrates the relative proportions of hypochlorous acid and hypochlorite ion as they vary with the pH of the solution (Sawyer 1978). When the information of Figure 4.2 is used in conjunction with the germicidal information of Figure 4.1, it can be seen that for water systems with a finished water pH of 7.5 or higher that more than 60 percent of any free chlorine present would be present as hypochlorite ion and the germicidal effectiveness of this solution approaches that of monochloramine. For water systems that use lime softening, stable waters are often achieved with a finished water pH of 9.2 or higher and for these systems the germicidal effectiveness of chloramines is very similar to one using free chlorine.

Blending

Blending, in the context of this report, refers to the practice of combining chlorinated water with chloraminated water. Although not a common practice, it is used primarily by those water systems that have both ground water and surface water as water supplies. In these cases, the surface supply is disinfected with chlorine or another oxidant with chloramines being used as the secondary disinfectant within the distribution system to avoid high DBP concentrations.

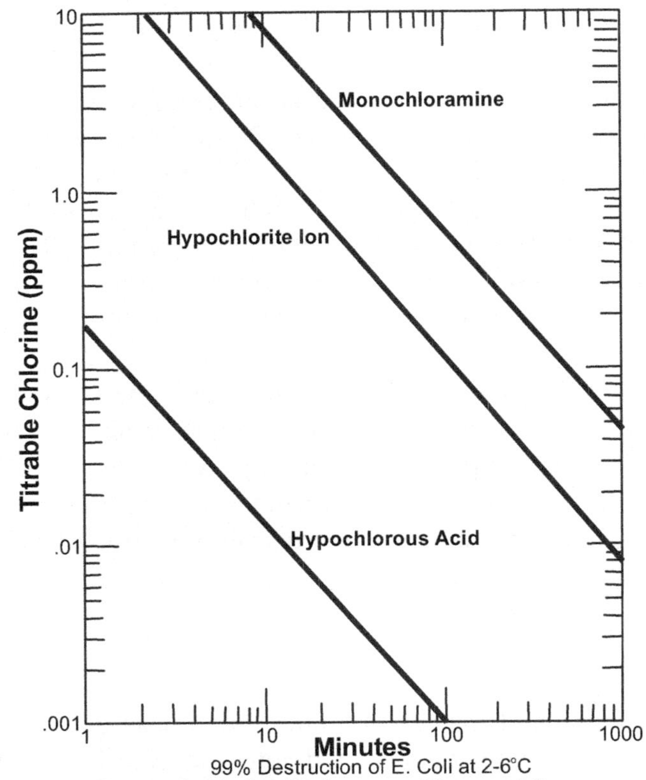

Source: White (1986)

Figure 4.1 Comparison of germicidal efficiency of hypochlorous acid, hypochlorite ion and monochloramine.

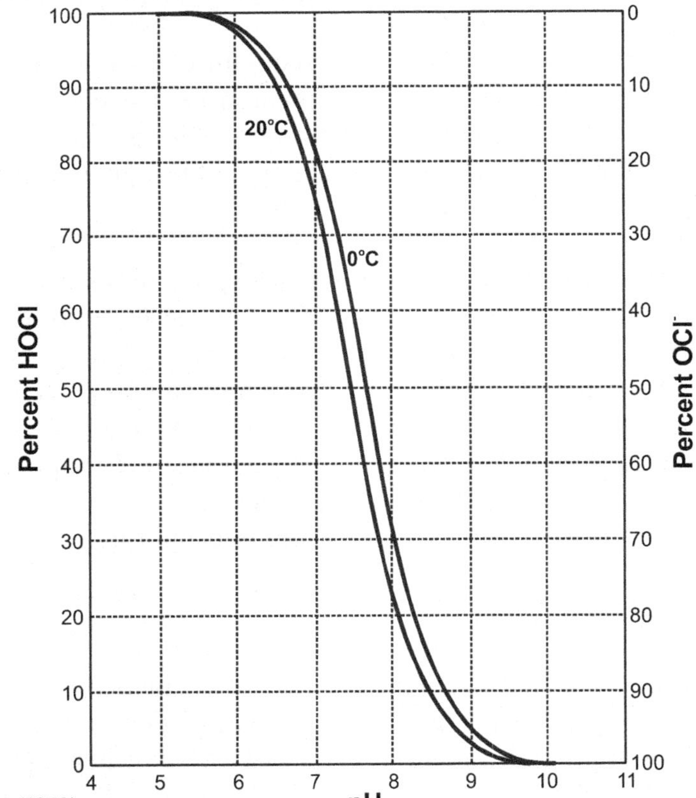

Source: Sawyer & McCarty (1978)

Figure 4.2 Effect of pH on the distribution of hypochlorous acid and hypochlorite ion in water.

Several wells are also located throughout the community and normally these ground waters receive only disinfection and possibly fluoridation prior to discharging directly into the distribution system. In areas where these two waters meet, changes in chlorine chemistry are certain to occur.

Of the 63 water systems surveyed, 21 systems used both ground water and surface water and 5 of these systems blend chlorinated and chloraminated water. Additionally, one system using surface water from two different sources also blends chlorinated and chloraminated water. There is not thought to be any implication regarding public health where blending occurs as long as a satisfactory chlorine residual is maintained, but rather the issue is mainly one of aesthetics.

Of the 6 utilities blending water, some do receive taste and odor complaints from areas where blending is likely to occur. As can be seen from Figure 3.1, if the chlorine:ammonia-nitrogen ratio should exceed about 5:1, chlorine residuals begin to decrease and the formation ofdichloramines is probable. These compounds do cause customer complaints by changing the taste and odor of the water.

One technique for using chlorinated ground water and chloraminated surface water within the same distribution system is to isolate one from the other through valves in the system. Often a particular pressure zone can be fed from either the ground or surface water source thus keeping the two waters completely separated from each other. Other systems are willing to accept a limited number of complaints where the two waters merge although the blending zone will vary daily depending upon local water use.

Some sophisticated techniques have been developed to deal with this problem. Metropolitan Water District of Southern California developed a computer model to allow the prediction of acceptable blends of chloraminated and chlorinated water without testing each possible blend (Barrett 1985). If the model predicted a chlorine:ammonia-nitrogen blend ratio of more than 5:1, additional testing was conducted. As noted previously, the most troublesome area for taste and odor complaints occurs when the chlorine:ammonia-nitrogen ratio exceeds 5:1.

SOCIAL ISSUES

The process by which a water supplier decides to implement chloramines as a secondary disinfectant is unique to each utility. The decision may be made at the staff level, management level or even at the political level. Much of the success will depend upon the utility structure, local government, public perception and communication. Who makes the decision and how the process is communicated can be pivotal to the success of the implementation. In some circumstances, utility staff will evaluate options and proceed with implementation. At the other end of the spectrum, a utility may bring in outside expertise to develop options, make recommendations, design and finally build the infrastructure. As unique as each case is, however, there are key factors which are common to successful efforts. Communication and coordination amongst the interested, affected, and potentially affected parties is paramount to a successful chloramine conversion. Understanding who your audience is at various stages of the process will also increase the opportunity for success.

No matter who the audience is, the first step is to address the need for a change. Most organizations (and people) are hesitant to embrace change. And, changes to their water supply are generally not readily accepted. In fact, when you announce a change to the water, in many cases you will pique an interest in the public that is out of the ordinary. This simple increase in customer interest alone is something for which a utility should plan at the onset of the project.

Announcing a change to "water treatment" will engender questions regarding water sources, existing treatment, costs, and more. Be prepared to provide adequate and clear information about all aspects of your utility. Various utility and AwwaRF reports have noted that communication is the lynchpin to the success (or failure) of a water project. Plan to communicate at all levels and plan to do it frequently.

It is important to develop a few key messages about what is driving the need for the change and why chloramines are the solution. A utility may be looking to resolve any number of challenges (such as new regulations, taste and odor, and/or disinfectant residual maintenance) through chloramines. It is important that the chloramines clearly represent a solution to your challenge and that your audience understand the nature of your problem. Otherwise, your customers may perceive that you are trying "to fix something that isn't broke." It may be incredibly clear to the utility why it is making this change; other agencies (or individuals) not involved in the business of water on a day to day basis may not readily grasp the pertinent circumstances. It is incumbent upon the utility to educate their audience.

It is important that you develop multiple avenues for communicating this package of information based on the audience. Communication with regulators should be different than the presentations made to the general public, the media or a utility governing board. Examples of newsletters, media releases, pamphlets and fliers are provided in the appendices. Successful programs involve regulators from the onset, inform political bodies early and often, and implement multiple avenues to inform the public.

An important audience that may be overlooked in planning a communication program for chloramines is utility staff. It is critical that internal staff also understand the "problem" and the solution. It is also important to make the internal audience as broad as possible, including field crews and customer service representatives as well as operators and engineers. Once the public becomes aware of the change, all persons that have any contact with a customer will become ambassadors of the change – even the field technician operating a hydrant may be approached by a curious customer. Providing internal staff with enough training and information will be extremely beneficial, for the chloramine conversion as well as for the utility in general. The utility employee who does not answer a customer question positively, or refer them to someone who can, may cause the customer to lose confidence in the utility. Using a "key message" format can streamline the staff training process as well as being effective with customers.

For a chloramine conversion or implementation, there are three key components to the message:

1. What is the challenge (regulations, taste, odor, etc.) and what is the solution?
2. Who will it affect? Are there any sensitive populations?
3. When will the solution be in place?

The public will have many questions; some will seem sophomoric, some will be outlandish. You will anticipate many of them, but there will be some that you would never imagine! It is important to answer each as a legitimate question.

Methods of Communication

Given the regional scope of the Tampa Bay Water conversion, the Public Information campaign was necessarily large, expansive and intensive. A smaller, more compact utility could be very successful with a less intense effort. The Communication Plan consisted of numerous components including:

1. Fact Pack (including Frequently Asked Questions and Answers, Glossary, Background White Papers, Special Populations Information)
2. Targeted Business Letters (Dialysis, pet stores)
3. Newsletters for Homeowners Association or local papers
4. Newspaper Inserts
5. Letters to the Editor
6. Radio Spots
7. Community Television Appearances
8. TV ads
9. Bill Inserts
10. Speakers Bureau
11. Call Center
12. Brochures
13. "Take one" Cards for Pet Stores
14. Pet Store Posters
15. Customer Service Workshops
16. Operator Training

Examples of these materials are provided in the appendices. The common theme of the materials was the key messages: what, who and when. The timing of various activities was coordinated amongst the utility, its consecutive systems, the regulatory agencies and the public relations consultant and planning began more than 12 months in advance of the conversion. Monthly update meetings were held until approximately 3 months before the conversion date after which the frequency was increased to biweekly and then weekly. At the time of the conversion, daily updates were organized.

General Notice to Hospitals and Dialysis Centers

A critical population that must be made aware of the chloramine conversion are dialysis centers. In the Tampa Bay area, the local organization that coordinates dialysis efforts, Network 7, informed the utility that dialysis centers are required to test their product water for both free chlorine and chloramines and are trained to do so routinely. A survey of local centers confirmed that the water used for dialysis is thoroughly evaluated for the presence of monochloramine and free chlorine as either could have detrimental effects on the dialysis process. The dialysis centers were also polled as to their preference of being the point of primary contact for their clients or having the utility send notice to individuals. Citing patient confidentiality and turnover in patients, the dialysis centers became the contact points for this population. A series of three letters were sent to each hospital and dialysis center beginning 6 months prior to the anticipated conversion date. A second letter was mailed 2 months before startup and a final notice went out one month before the implementation. The final letter requested a response be returned to the utility acknowledging the conversion date. Follow-up phone calls also documented complete coverage of the affected medical community. No adverse incidents were reported and Tampa

Bay Water received a community service award from the dialysis community in recognition of the communication program.

One hospital employee called inquiring if the conversion would affect housekeeping, janitorial services or laundry. He was informed that there are no known impacts and was also provided with contacts at other hospitals in the region which already had a chloraminated water supply. No adverse incidents have been reported from hospitals regarding these activities.

Other Health Issues

A number of inquiries were received on the following topics: contact lenses, immune deficiencies, allergies, dermal absorption of chlorine and chloramines, ulcers and digestive disorders. A review of literature and inquiries to national experts found no indications that chloramines were related to or had any impacts on these topics. Customers were directed to contact their health care providers in all cases for specific concerns and recommendations. A number of customers wanted examples of "original research". In these cases, references to USEPA documents available on the website were provided. In a small number of cases, customers were not satisfied with this response; in all cases, customers were provided with general information regarding the use of chloramines throughout the country and over the past 4 decades or more. Understanding that chloramines are not a "new technology" and that the region was not being used as a "guinea pig" was extremely helpful to customers.

Utility staff are committed to tracking research and publications regarding chloramines (as well as other oxidants) and their potential health effects. Emphasizing this to customers was reassuring also.

Fish

Fish, both saltwater and freshwater species, are very sensitive to both free chlorine and chloramines and may die if exposed to these oxidants. The challenge with chloramines is two-fold: they do not dissipate rapidly so letting the water sit for a day or two will not make it safe for fish and the "chlorine neutralizers" are not effective on chloramines. Chloramine neutralizing chemicals are available in pet stores and a reconnaissance of several stores prior to the conversion confirmed that both were available even in the region not utilizing chloramines.

Pet stores, department stores, and feed stores (contact point for outdoor ponds) were approached and asked or permission to post posters and "take one" card displays. (Examples provided in Appendix.) All contacts were positive and were provided the materials by the utility. Follow-up visits and phone contacts were provided to ensure that the store did not run out of the cards. In total, 7 "fish kills" were reported to the utility, four of them in outdoor ponds. One pet store reported a large loss of fish approximately three days into the conversion, but no clear connection to the conversion could be discerned. Evaluation of the circumstances indicated that "out of date" dechloramination chemicals may have been inadvertently used.

Consideration was also made to contact fish markets, supermarkets, bait stores, and restaurants that might keep live fish on their premises. Letters were distributed to these businesses informing them of the switch to chloramines (including the why, when and where from the key messages), explaining how to dechloraminate the water and recommending that a pet supply vendor be contacted for further information if desired.

During a "Post conversion lessons learned workshop", it was noted that nurseries were not directly contacted for notification. This business segment caters to outdoor pond enthusiasts and would have served as another avenue to communicate with these customers. It was also learned that there is a state organization of nurserymen and growers that could have served as a conduit for information.

Pets

Cats, dogs, and birds were specifically addressed in a "Frequently Asked Questions and Answers" document provided by mail or on the utility website. A call center was activated for one month around the conversion and also provided with a script to answer such inquiries. Several veterinarians in a local chloraminated system were interviewed about impacts to pets and no issues or concerns were identified by these professionals. In addition to calls regarding more common pets, customers also called about monkeys, reptiles, insects, and amphibians. With the exception of one reptile group (turtles) and amphibians, no known adverse effects are reported in the literature for exposure to or consumption of chloraminated water. Turtles and amphibians spend a significant amount of time in water and, based on recommendations of a local zoological garden, it was recommended that both chlorine and chloramines be removed from their water. No adverse impacts to any pets have been reported to the utility. An interesting aspect of this effort was learning of the diversity of pets in the community.

Household concerns

A number of concerns may arise regarding routine household activities. Pool maintenance, cleaning surfaces, icemakers, watering plants and laundry were the most common "household inquiries" encountered in the Tampa Bay conversion. An interesting comment from several customers was consideration of the safety of mixing ammonia and chlorine. For example, the following inquiry was made via email:

> Why are you using a combination of two chemicals (bleach and ammonia) which, according to both labels, appears to be a very dangerous combination. Here is label information:
>
> "Bleach contains the strong oxidizer Sodium Hypochlorite. Be sure to always flush drains thoroughly before and after use. Do not use or mix with other chemicals such as ammonia, toilet bowl cleaners, acids or other household chemicals. To do so will cause the release of toxic gases."

In response, the customer was assured that the levels of ammonia and bleach used in the process of making chloramines was orders of magnitude lower than what is found in the bottles under the kitchen sink. Also, trained, licensed operators perform this activity under controlled conditions in a water treatment plant. For most customers, this explanation sufficed.

Similarly, the low levels of ammonia and chlorine used to form chloramines are miniscule in comparison to chemical concentration of products used for household cleaning and pool maintenance; special considerations are not warranted. Local nurseries confirmed that the low levels of ammonia that could be present in the water supply should not pose any problems for houseplants, gardens or "garden friendly" insects.

Several customers reported that following the conversion, their icemakers began to produce cloudy ice cubes. No explanation has been provided for these limited number of observations. Another unexplained observation was an increase in the number of "green water" complaints received by the local utilities. Some possible explanations have been suggested including incomplete oxidation of color which is normally present in local groundwater and possible chemical interactions in newly installed plastic piping. Sample collection and analyses has not confirmed the occurrence of green water or chemicals of concern to date. More investigation of this phenomenon is warranted, however, due to the consistency in reports received.

Point of Use Devices

The most frequent inquiry throughout the public information campaign revolved around point of use devices: water filters and water softeners. An unexpected consequence of the conversion was a dramatic increase in sales efforts for point of use devices. Sales pitches ranged from pressuring old customers to "upgrade" their filters or softeners to actual misrepresentation of media material in a manner that suggested the drinking water supply was no safe. In cases where these tactics could be documented, the companies involved were notified by legal counsel to "cease and desist" such activities. Also, several local municipalities distributed "Customer Alert Notices" in their water bills regarding unethical POU sales tactics. Two utilities developed special television segments to be aired on the government access channel.

CHAPTER 5
CHLORAMINE SURVEY RESULTS

The chloramination survey form (Appendix B) was mailed to 111 U.S. utilities. In the survey, utilities were asked to provide information about their treatment process, treatment capacity, source water, problems during startup, experience using chloramines, problems and concerns, and current chloramination practices. Responses were received from 63 utilities. The participating utilities represent a broad cross-section of geography, plant size, and experience using chloramines.

Follow-up telephone interviews were conducted with utilities to clarify any ambiguities in the survey responses. Site visits were made to 13 of the utilities to conduct in-depth face-to-face interviews. The results of the survey and the interviews are reported in this chapter.

REPSONSES TO THE WRITTEN SURVEY

The survey data were stored and organized in an Access database, and this database was queried to tabulate and summarize the information. The survey results illustrate the state of chloramination practice in the U.S. and report the experience, opinions, and concerns of the water treatment professionals who practice chloramination daily.

Why Do You Use Chloramines?

Utilities were asked if they use chloramines to reduce DBP levels, improve taste, or control biofilms. Each utility was asked to check all of the above that apply, and if there are other reasons, to provide comment. The utilities' responses are summarized in Figure 5.1. The biggest reason for using chloramines is to reduce DBP levels, selected by over 76 percent of the utilities. The reason most frequently volunteered by utilities under the "other" heading was "to maintain a residual", written in by 14 utilities.

Utilities that use chloramines to reduce DBPs were asked to compare their maximum and average TTHM concentrations with chlorine to those after switching to chloramines. Twenty-eight utilities provided this comparison data. The maximum TTHMs formed using chlorine and chloramines and the reduction in TTHMs after switching to chloramines are summarized in Table 5.1.

Table 5.1
Comparison of TTHM formation with chlorine and chloramines

Maximum TTHMs Using Chorine (Mean)	181 μg/L
Maximum TTHMs Using Chloramines (Mean)	24 μg/L
Reduction in TTHMs with Chloramines (Mean)	82 %

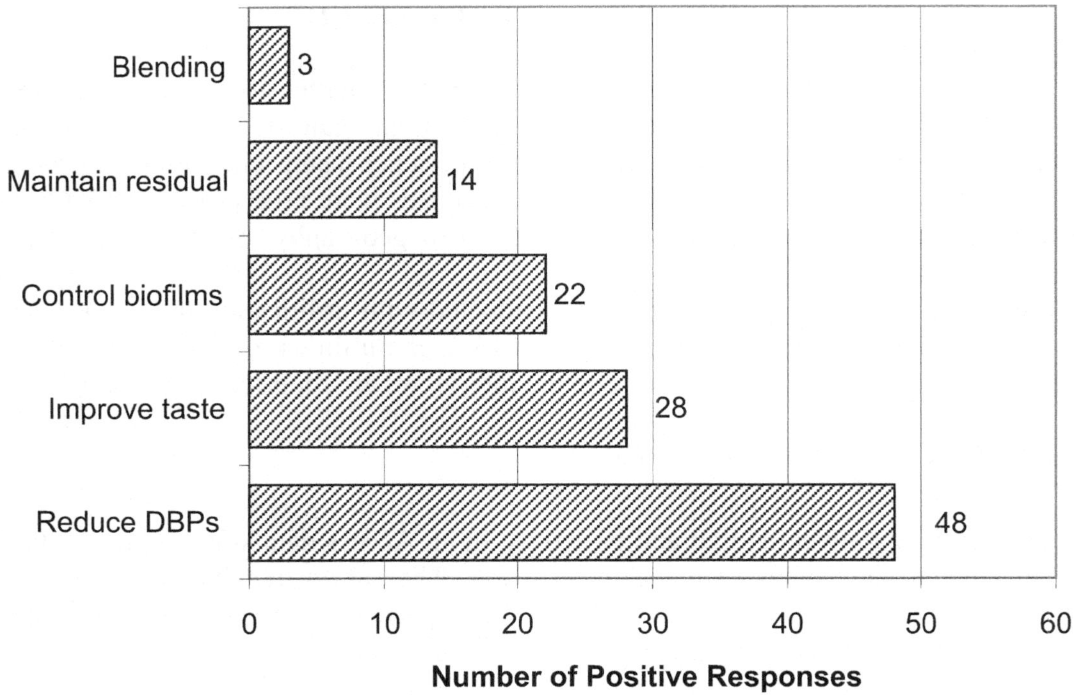

Figure 5.1 Reasons stated for using chloramines

How Long Have You Used Chloramines?

There was a broad range of experience with chloramination among the surveyed utilities, as illustrated by Figure 5.2. The shortest period of use was 1 year, and the longest was 75 years. The median for chloramination experience was 20 years.

Water Source and Treatment Capacity

The number of utilities treating surface water, groundwater, or a blend of the two is shown in Table 5.2. Their total treatment capacities are summarized in Table 5.3.

Table 5.2
Source water treated

Source Water	Number Treated
Surface Water	49
Groundwater	3
Both Surface Water and Groundwater	11

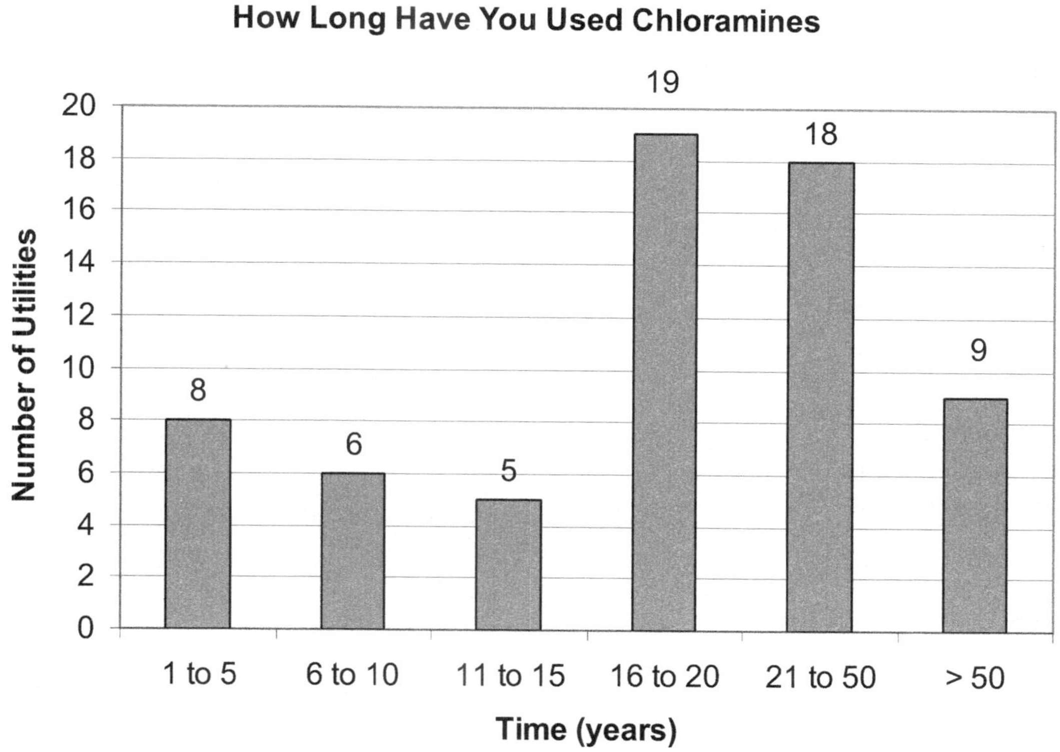

Figure 5.2 Length of time using chloramines

Table 5.3
Total treatment capacity

Treatment Capacity (mgd)	Number of Utilities*
0 to 50	22
51 to 100	15
101 to 200	10
> 200	11

* Four utilities left this item blank and one purchases water.

Problems During Startup

Utilities were asked to describe any significant startup problems, either at the plant or in the distribution system. Eleven utilities stated they had problems during startup as summarized in Table 5.4. Of those that discussed the resolution of their problem, all but one found a solution quickly. The utility that did not find a solution quickly had a problem obtaining their target residual in the distribution system and worked on this issue for two years before resolving the problem. Of the two utilities that had taste and odor problems, one attributed the problem to chloramines (dichloramine formation) and the other to taste and odor compounds that were not adequately oxidized by chloramines. The utilities that had problems with fish kills or at the dialysis center alerted these customers about the potential impact of chloramines but the customers failed to prepare adequately.

Table 5.4
Problems during startup

Problem	Number of Utilities
Fish Kills	2
Getting target residual	2
Feeding ammonia	2
Taste and odor	2
Dialysis centers	1
Increased iron levels	1
Bacterial growth	

Periodic Use of Free Chlorine

Utilities were asked if they sometimes use short periods of free chlorine instead of chloramines for disinfection within the distribution system. Fourteen of 63 utilities (22 percent) switch to free chlorine periodically. Those that do were asked to describe their practices during the switch, their reason for switching, and any documented benefits. Their responses are summarized in Tables 5.5, 5.6, 5.7, and 5.8.

The most common reasons given for switching to free chlorine were to reduce biofilms and to control nitrification, listed by four utilities. Three utilities use free chlorine as part of their routine flushing program. All but two of the 14 utilities switch to free chlorine once a year. One utility switches every 6 months and the other biannually.

The duration of free chlorine use ranged from 1 week to 3 months, but most utilities switch for either 1 to 2 weeks or 4 to 6 weeks. Twelve of the 14 utilities notify their industrial and medical customers of the switch, but only five notify the public. Methods for notifying their customers include certified mail, telephone, newspapers, television, radio, faxes, and the Internet. Ten of the 14 utilities notify their state agency. Five of the utilities monitor DBPs during free chlorine use, and four stated that their State agency suspends DBP sampling and monitoring requirements during the switch.

The time utilities spend preparing for the switch to free chlorine ranges between 2 hours and 40 hours with a median time of 9 hours. The utilities use one to 10 staff to implement the switch, and the median is 3 people. Eight of the utilities claimed documented benefits from the periodic use of free chlorine. All but one of the utilities stated control of nitrification, maintenance of residual, or both as the benefit it derives. The utility that did not list nitrification control or residual maintenance claimed a reduction in coliform occurrences.

Table 5.5
Why do you switch to free chlorine?

Reason	Number of Utilities
Reduce biofilms	4
Control nitrification	4
Part of routine flushing program	3
DBP control	1
Maintain CT while using PAC	1
No reason stated	1

Table 5.6
How often do you switch to free chlorine?

Frequency	Number of Utilities
Twice a year	1
Annually	12
Every two years	1

Table 5.7
How long do you use free chlorine?

Duration	Number of Utilities
1 to 2 weeks	6
2 to 4 weeks	2
4 to 6 weeks	5
3 months	1

Table 5.8
Practices during free chlorine use

Practice	Yes	No
Are industrial and medical customers notified?	12	2
Is the public notified?	5	9
Is your State agency notified?	10	4
Do you flush during the switchover?	5	9
Do you do additional sampling during the switchover?	6	8
Do you measure DBPs while using free chlorine?	5	9
Are customer complaints or reactions monitored?	12	2
Do you have any documented benefits of the switchover?	8	6

Chloramination Practices

Utilities were asked qualitative and quantitative questions about their chloramination practices. Qualitative responses are summarized in Table 5.8. Statistical summaries of the qualitative data supplied by the 63 utilities are shown in Table 5.9. Approximately 11 percent of the utilities stated they have difficulty maintaining their target Cl:NH_3-N ratio. The length of experience of these 11 utilities ranged from 10 to 35 years. The median Cl:NH_3-N ratio was 4 and values ranged between 3 and 6.5.

The mean distribution system residence time is 2.6 days. Eight percent of the utilities blend chlorinated and chloraminated water in their systems, and 22 percent practice booster chlorination or chloramination. Seventy-nine percent of the utilities add chlorine ahead of ammonia. The median time between chlorine and ammonia addition for these utilities is 25 minutes, and lag times range between less than a minute to 6 hours. The median pH at the ammonia addition point is 7.5. Total chlorine residuals at the ends of distribution systems are between 0.03 mg/L and 3.0 mg/L, with a median residual of 1.0 mg/L.

Eight utilities make seasonal water quality monitoring changes, with increased monitoring for signs of nitrification being the most frequently cited change (four utilities). Two utilities mentioned increased monitoring of residuals, and one adjusts dose for changes in the raw water blend. The remaining utility increases HPC monitoring seasonally.

Table 5.9
Practices during chloramination, qualitative responses

Question	Yes	No
Do you have difficulty maintaining your target Cl:NH_3-N ratio?	7	56
Do you apply free chlorine ahead of ammonia?	50	13
Do you blend chlorinated water with chloraminated water?	5	58
Do you have any current difficulties using chloramines?	10	53
Do you practice booster chlorination or chloramination?	14	49
Do you provide information on residuals to consumers?	33	30
Do you make seasonal changes to water quality monitoring?	8	55

Table 5.10
Practices during chloramination, quantitative responses

Practice	Mean	Median	Minimum	Maximum
Cl:NH_3-N ratio	4.3	4	3	6.5
Average distribution system residence time (days)	2.6	2	0.1	21
Maximum distribution system residence time (days)	5.5	4	1	28
Time between chlorine and ammonia added (min)	83	25	0	720
pH at the ammonia addition point	7.9	7.5	6.4	9.8
Residual at the end of the distribution system (mg/L)	1.3	1.0	0.03	3.0

Problems and Concerns with Chloramination

Utilities were asked about their problems and concerns with chloramination in three survey questions. First, they were asked if they had any problems or concerns with the storage, mixing, or feeding of chlorine and ammonia. Thirty percent of the utilities have problems or concerns in this area. The major issues of concern are safety in handling and storing the hazardous chemicals and difficulty feeding the chemicals.

In a subsequent question, 19 issues commonly described as concerns with chloramination were listed and utilities were asked to mark all that were either a concern or a problem. The responses to this question are tabulated in Table 5.11. The biggest concern utilities have about chloramination is nitrification. Nitrification was also most frequently cited as a problem along with maintaining system residuals.

When asked if they were currently experiencing problems with chloramination, 16 percent of the utilities responded "yes". Feeding ammonia was the most frequently cited problem (by three of the utilities) and other problems include nitrification, maintaining a residual, disinfecting new mains, and mixing.

Table 5.11
Problems and concerns using chloramines

Issue	Concern	Problem
Nitrification	34	7
Biofilms	20	2
Medical issues	17	0
Disinfection efficiency	15	0
Aquariums	14	4
Free ammonia at some locations	14	4
Blending	14	3
Maintaining system residuals	13	7
Fish ponds	13	3
Increase in microorganisms	10	0
Need to flush pipes	9	1
Degradation of distribution system materials	8	4
Taste and odor	8	3
Blending with wholesale customers	8	0
Using ammonia	8	0
Operations	5	1
Customer acceptance	5	0
Corrosion	4	1
Uses more chlorine	2	0

REFERENCES

AWWA. Chlorine-Ammonia Treatment. 1941. Joint Committee Report. *Jour. AWWA*, 33:12:2079.

AWWA Disinfection Systems Committee. 2000. Committee Report: Disinfection at Large and Medium Size Systems. *Jour. AWWA,* 92:5:32.

Baker, R.J. 1959. Types and Significance of Chlorine Residuals. *Jour. AWWA*, 51:9:1185.

Barrett, S.E., M.K. Davis, and M.J. McGuire. 1985. Blending Chloraminated and Chlorinated Waters. *Jour. AWWA*, 78:1:50.

Bellar, T.A, J.J.Lichtenberg, and R.C. Kroner. 1974. The Occurrence of Organohalides in Chlorinated Drinking Water. *Jour. AWWA,* 66:12:703.

Bishop, M.M. Survey of Water Disinfection Practices. 1992. *Jour. AWWA,* 84:9:121.

Braidech, M.M. 1931. Cleveland Uses Ammonia-chlorine to Check Phenol Tastes in Water. *Water Works Engrg*. 84:433.

Brodtmann, N.V.and P.J. Russo. 1979. The Use of Chloramines for Reduction of Trihalomethanes and Disinfection of Drinking Water. *Jour. AWWA,* 71:1:40.

CA Department of Health Services. 2002. Studies on the Occurrence of NDMA in Drinking Water. [Online]. Available: <http://www.dhs.ca.gov/ps/ddwem/chemicals/NDMA/studies.htm >[cited May 1, 2003].

Caterson, R.J., E. Savdie, E. Raik, D. Couts, and J.F.Mahony. 1982. Heinz-body Hemolysis in Hemodialyzed Patients Caused by Chloramines in Sydney Tap Water. *The Medical Journal of Australia,* October 16, 1982.

Choi, J. and R.L.Valentine. 2002. Formation of N-nitrosodimethylamine (NDMA) from reaction of monochloramine: a new disinfection by-product. *Water Research*, Vol. 36, No. 4, pp. 817-824.

Choi, J., S.E.Duirk, and R.L.Valentine. 2002. Mechanistic Studies of N-nitrosodimethylamine formation in Drinking Water. *Journal of Environmental Monitoring*, 4(2), 249-252.

Choi, J., and R.L. Valentine. 2002. A kinetic model of N-nitrosodimethylamine (NDMA) formation during water chlorination/chloramination. *Journal Water Science and Technology*, Vol. 46, No. 3, pp. 65-71.

Clarke, N.A., G. Berg, P.W. Kabler, and S.L. Chang. 1962. Human Enteric Viruses in Water: Source, Survival and Removability. *International Conf. Water Pollution Research*, London: Pergamon Press.

Collins, H. and R. Selleck. 1972. Process and Kinetics of Wastewater Chlorination. SERL Rept. 72-5, Univ. of Calif. Berkley.

Dennis, J.F., D.C. Rauscher, D.A. Foust. 1991. Practical Aspects of Implementing Chloramines. Proc. AWWA Annual Conf., Denver, CO. AWWA.

DiGiano, F.A.; D.E. Francisco, W. Zhang, and L. Todd. 2000. Bacterial Regrowth in Drinking Water Distribution Systems: A Comparison of Durham and Raleigh. *Water Resources Research Institute.* Report No. 326.

Dufor, C.N. and E. Becker. 1962. Public Water Supplies in the 100 Largest Cities in the United States. USGS Water Supply Paper 1812. USGS, Washington, D.C.

Eaton, A. and M. Briggs. 2000. NDMA: Analysis of a New DBP. In *Proc. 2000 AWWA WQTC,* Salt Lake City, Utah.

Ely, H.M. 1933. Ammonia-Chlorine Treatment in Danville, Illinois. *Jour. AWWA*, 25:9:1257.

Feng, T.H. Behavior of Organic Chloramines in Disinfection. 1966. *Jour. WPCF,* 38:4:614.

Faust, S.D., and O.M. Aly. 1998. *Chemistry of Water Treatment*, 2nd Ed., Ann Arbor Press, Chelsea, MI

Goehring, E.G. 1931. Ammonia-chlorine Treatment at Beaver Falls and New Brighton, Pa. *Jour. AWWA*, 23:9:1372.

Hoehn, R.C. and D.C. Johnson. 1983. Disinfection, Water Quality Control, and Safety Practices of the Water Utility Industry in 1978 in the United States. *Jour. AWWA,* 75:1:51.

Jafvert, C. and R.L. Valentine. 1992. A Reaction Scheme for Chlorination of Ammoniacal Water. *Environmental Science and Technology*, Vol. 26, No. 3, pp. 577-586.

KS Department of Health and Environment. 1995. Policies, General considerations and Design Requirements for Public Water Supply Systems in Kansas. Bureau of Water, Topeka.

Kimura, E.T., D.N. Ebert, and P.W. Dodge. 1971. Acute Toxicity and Limits of Solvent Residue for Sixteen Organic Solvents. *Toxicol. & Appl. Pharmacol.* 19:699.

Kirmeyer, G.J., G.W. Foust, G.L. Pierson, J.J. Simmler, and M.W. LeChevallier. 1993. *Optimizing Chloramine Treatment.* AWWA Research Foundation and AWWA. Denver, Colorado.

Kirmeyer, G.J., L.H. Odell, J. Jacangelo, A. Wilczak, and R. Wolfe. 1995. *Nitrification Occurrence and Control in Chloraminated Water Systems.* AWWA Research Foundation and AWWA. Denver, Colorado.

Kjellstrand, C.M., J.W. Eaton, and Y. Yawata. 1974. Hemolysis in Dialyzed Patients Caused by Chloramines. *Nephron.* 13: 427-433.

Klein, E. 1986. Effects of Disinfectants in Renal Dialysis Patients. *Environmental Health Perspectives,* Vol. 69, pp. 45-47.

Kreft, P., M. Umphres, J. Hand, C. Tate, M.J. McGuire, and R.R. Trussell. 1985. Converting from Chlorine to Chloramines: A Case Study. *Jour. AWWA*, 78:1:38.

Kruse, C.W., M.C.Snead, and V.P. Olivieri. 1981. Design of Alternative Disinfecting Systems with Special Reference to Water Quality at the Consumers Tap. In *Proc. 23rd Public Wtr. Supply Engrg. Conf.*, Univ. of Illinois, Urbana.

Kuzma, R.J., D.M. Kuzma, and C.R. Buncher. 1977. Ohio Drinking Water Source and Cancer Rates. *Amer. Jour. Public Health*, 67:8:725.

LeChevallier, M.W., T.M. Babcock, and R.G. Lee. 1987. Examination and Characterization of Distribution System Biofilms. *Appl. & Envir. Microbiol.,* 53:12:2714.

LeChevallier, M.W., N.J. Shaw, and D.B. Smith. 1996. *Factors Limiting Microbial Growth in Distribution Systems: Full-Scale Experiments.* AWWA Research Foundation and AWWA. Denver, Colorado.

Lieu, N.I., R.L. Wolfe, and E.G. Means III. 1993. Optimizing Chloramine Disinfection for the Control of Nitrification. *Jour. AWWA*, 86:2:84

Lyles, J.E. 1931. Tastes and Odors in Small water Supply Removed by Ammonia. *Water Works Engrg.* 84:13:1003.

Margerum, D.W., E.T. Gray, and R.P. Huffman. 1978. Chlorination and the Formation of N-Chloro Compounds in Water Treatment. *Organometals and Organometaloids: Occurrence and Fate in the Environment.* ACS, Washington, D.C.

McGuire, M.J., N.I. Lieu, and M.S. Pearthree. 1999. Using Chlorite Ion to Control Nitrification. *Jour. AWWA*, 91:10:52.

Mitcham, R.P., M.W. Shelley, and C.M. Wheadon. 1983. Free Chlorine Versus Ammonia-chlorine: Disinfection, Trihalomethane Formation, and Zooplankton Removal. *Jour. AWWA,* 75:4:196.

Morris, J.C. and R.A. Isaac. 1983. Review of Kinetic and Thermodynamic Constants for the Aqueous Chlorine-ammonia System. *Water Chlorination: Environmental Impact and Health Effects.* Vol. 4. Ann Arbor Sci. Publ., Inc. Ann Arbor, Mich.

Najm, I. and R.R. Trussell. 2001. NDMA Formation in Water and Wastewater. *Jour. AWWA,* 93:2:92.

Norman, T.S., L.L. Harms, and R.W. Looyenga. 1980. The Use of Chloramines to Prevent Trihalomethane Formation. *Jour. AWWA,* 72:3:176.

Norton, C.D. and M.W. LeChevallier. 1997. Chloramination: Its Effect on Distribution System Water Quality. *Jour. AWWA,* 89:7:66.

Ozekin, K., R.L. Valentine, and P.J. Vikesland. 1996. Modeling Chloramine Decay in Natural Waters. American Chemical Society (ACS) Symposium Series 649, *Water Disinfection and Natural Organic Matter*, pp. 113-125.

Page, T., R.H. Harris, and S.S. Epstein. 1976. Drinking Water and Cancer Mortality in Louisiana. *Science*. 193:55.

Regan, J.M., G.W. Harrington, and D.R. Noguera. 2002. Ammonia- and Nitrite-Oxidizing Bacterial Communities in a Pilot-scale Chloraminated Drinking Water Distribution System. *Applied and Environmental Microbiology*. Jan. 2002, p. 73-81.

Reiber, S. 1993. Investigating the Effects of Chloramines on Elastomer Degradation. *Jour. AWWA*, 85:8:101.

Reilly, J.K. and J.S. Kippin. 1983. Relationship of Bacterial Counts with Turbidity and Free Chlorine in Two Distribution Systems. *Jour. AWWA,* 75:6:309.

Rook, J.J.. 1974. Formation of Haloforms During Chlorination of Natural Waters. 1974. *Water Trmnt. Exam.,* 23:234.

Sawyer, C.N., and P.L. McCarty. 1978. *Chemistry for Environmental Engineering*, 3rd Ed. McGraw-Hill Book Co., New York.

Shull, K.E. 1981. Experience with Chloramines as Primary Disinfectants. *Jour. AWWA*, 73:2:104.

Spaulding, C.H. 1929. Preammonation at Springfield, Illinois. *Jour. AWWA*, 21(8):1085.

Symons, J.M., R. Xia, G.E. Speitel Jr., A.C. Diehl, C.J. Hwang, S.W. Krasner, and S.E. Barrett. 1998. *Factors Affecting Disinfection By-Product Formation During Chloramination.* Denver, CO. AWWARF and AWWA.

Tracy, H.W., V.M. Camarena, and F. Wing. 1966. Coliform Persistence in Highly Chlorinated Waters. *Jour. AWWA,* 58:9:1151.

U.S. Environmental Protection Agency. 1978. National Interim Primary Drinking Water Regulations. Control of Organic Chemical Contaminants in Drinking Water. *Fed. Reg.*, 43:5756.

U.S. Environmental Protection Agency. 1979. National Interim Primary Drinking Water Regulations. Control of Trihalomethanes in Drinking Water: Final Rule. *Fed. Reg.*, 44:68624.

U.S. Environmental Protection Agency. 1981. Treatment Techniques for Controlling Trihalomethanes in Drinking Water. Drinking Water Research Div., Office of Research and Development, Cincinnati, OH.

U.S. Environmental Protection Agency. 1989. 40 CFR Parts 141 and 142. National Primary Drinking Water Regulations; Filtration, Disinfection; Turbidity, Giardia lamblia, Viruses, Legionella, and Heterotrophic Bacteria; Final Rule. *Fed. Reg.* (Part II), 54(124):27486-27541.

U.S. Environmental Protection Agency. 1991. Guidance Manual for Compliance with the Filtration and Disinfection Requirements for Public Water Systems Using Surface Water Sources.

U.S. Environmental Protection Agency. 1998. Drinking Water Contaminant Candidate List. *Fed. Reg.* (Part III), 63(40):10274-10287.

Valentine, R.L. 1986. Bromochloramine Oxidation of N,N-Diethyl-p-Phenylenediamine in the Presence of Monochloramine. *Environ. Sci. & Technol.* 20(2):166-170.

Valentine, R.L., and C.T. Jafvert. 1988. General Acid Catalysis of Monochloramine Disproportionation. *Environmental Science & Technology*. 22:6:691-696.

Vikesland, P., K. Ozekin, and R.L. Valentine. 1998. Effect of Natural Organic Matter on Monochloramine Decomposition: Pathway Elucidation Through the Use of Mass and Redox Balances. *Environ. Sci. & Technol.* 32(10):1409-1416.

Vikesland, P., K. Ozekin, and R.L. Valentine. 2001. Monochloramine Decay in Model and Distribution System Water. *Water Research* , Vol. 35, no. 7, pp. 1766-1776.

Volk, C.J. and M.W. LeChevallier. 2000. Assessing Biodegradable Organic Matter. *Jour. AWWA,* 92:5:64.

Wajon, J.E., and J.C. Morris. 1980. Bromamination Chemistry: Rates of Formation of CH_2Br and Some N-Bromamino Acids. In *Water Chlorination: Chemistry, Environmental Impacts and Health Effiects.* Vol. 3. Edited by R.L. Jolley, W.A. Brungs, R.B. Cumming, and V.A. Jacobs. Ann Arbor, MI: Ann Arbor Science Publishers.

White, G.C. 1986. *The Handbook of Chorination*, 2nd Ed. Van Nostrand Reinhold Co., New York.

Wierenga, J.T. 1985. Recovery of Coliforms in the Presence of a Free Chlorine Residual. *Jour. AWWA,* 77:11:83.

Wilczak, A., J.G. Jacangelo, J.P. Marcinko, L.H. Odell, G.J. Kirmeyer, and R.L. Wolfe. 1996. Occurrence of Nitrification in Chloraminated Distribution Systems. *Jour. AWWA*, 88:7:74.

Wolfe, R.L., N.I. Lieu, G. Izaguirre, and E.G. Means. 1990. Ammonia-Oxidizing Bacteria in Chloraminated Distribution System: Seasonal Occurrence, Distribution, and Disinfectant Resistance. *Applied and Environmental Microbiology.* Feb. 1990, p. 451-462.

Wolfe, R.I., N.R. Ward., and B.H. Olson. 1984. Inorganic Chloramines as Drinking Water Disinfectants: A Review. *Jour. AWWA*, 76(5):74-88.

World Health Organization. 1993. *Guidelines for Drinking Water Quality*, 2nd Ed.. [Online]. Available: <http://www.who.int/docstore/water_sanitation_health/GDWQ/Chemicals/cyanogenchloridesum.htm >[cited May 1, 2003].

ABBREVIATIONS

AOC	assimilable organic carbon
AWWA	American Water Works Association
AwwaRF	Awwa Research Foundation
CDHS	California Department of Health Services
Cl_2:N or Cl_2:NH_3N	chlorine to ammonia-nitrogen ratio
cfu/mL	colony forming units per milliliter
°C	degrees Celsius
CNX	cyanogen halide
DBP	disinfection byproduct
D/DBP	disinfectant/disinfection byproduct
DBPR	Stage 1 Disinfectants and Disinfection Byproducts Rule
DOX	dissolved organic halogen
DO	dissolved oxygen
EPA	Environmental Protection Agency
FPA	flavor profile analysis
g	gram
HAA5	sum of the mass concentrations of five commonly measured haloacetic acids
HPC	heterotrophic plate count
L	liter
µg/L	microgram per liter
MCL	maximum contaminant level
MCLG	maximum contaminant level goal
mg/L	milligram per liter
M	molar
mol	mole
MPN	most probable number
ng/L	nanogram per liter
NMDA	N-nitrosodimethylamine
NOB	nitrite oxidizing bacteria
pH	negative log of the effective hydrogen ion concentration
s	second
SDS	simulated distribution system
SWTR	Surface Water Treatment Rule
T	temperature
THMs	trihalomethanes
TTHMs	total trihalomethanes
TOC	total organic carbon
USEPA	United States Environmental Protection Agency
VOC	volatile organic compound
WTP	water treatment plant
WHO	World Health Organization

APPENDIX A
LITERATURE REVIEW

HISTORICAL PERSPECTIVE

Chloramines have been used in the U.S. for drinking water disinfection for over 70 years. Chloramine use was prevalent in the 1930's (Spaulding 1929, Braidech 1931, Lyles 1931, Goehring 1931, and Ely 1933). A nationwide survey conducted in 1938 showed that 16 percent of municipal utilities used chloramines at some point in their treatment process (AWWA 1941). Chloramine use declined during WWII because of the inability to obtain ammonia (White 1986), and by 1959, the number of utilities using chloramines had decreased to 6 percent (Baker 1959). By the early 1960's, chloramine use had dropped to less than 3 percent (Durfor and Becker 1962).

In the early 1970s, researchers discovered that halogenated organics were formed during chlorination (Rook 1974, Bellar and Lichtenberg 1974). Subsequent toxicological studies classified some of these disinfection byproducts (DBPs) as possible human carcinogens (Kimura Ebert and Dodge 1971, Page, Harris, and Epstein 1976, Kuzma, Kuzma, and Buncher 1977). There has been resurgence in interest in chloramination over the last 30 years as utilities seek a means to provide secondary disinfection while limiting disinfection byproduct (DBP) formation. In 1979, EPA published the first regulations limiting DBPs in the National Interim Primary Drinking Water Regulations for Control of Trihalomethanes. A maximum contaminant level (MCL) for total trihalomethanes (TTHMs) was set at 100 μg/L. As researchers began to evaluate alternative disinfectants to limit DBP formation, it was discovered that chloramines resulted in significantly less TTHM formation than free chlorine (Brodtman and Russo 1979 and Norman, Harms, and Looyenga 1980).

Laboratory results, in which chloramines required much higher concentrations and longer contact time to achieve microbiological inaction levels comparable to chlorine, led EPA to impose a ban on use of chloramines (National Interim Primary Drinking Water Regulations, USEPA 1978). Field results, however, indicated a much higher chloramine biocidal efficiency than predicted by laboratory research (Shull 1981, Brodtman and Russo 1979, and Mitcham, Shelley, and Wheadon 1983). These successful applications of chloramination in field tests convinced EPA to rescind the ban a year later (USEPA 1979). The discrepancy between lab and field results for chloramine biocidal efficiency has been theorized to result from the relative resistance of in situ and laboratory grown organisms, the method of chloramine application (lab tests were typically conducted with preformed chloramines), and the criteria for evaluating the effectiveness of a disinfectant.

The Disinfection Systems Committee of AWWA conducted a survey of disinfection practices of large and medium sized utilities (serving more than 10,000 people) in 1998 (AWWA Disinfection Systems Committee 2000). This survey extended the database collected in similar surveys conducted in 1979 (Hoehn and Johnson 1983) and 1989 (Bishop 1992). The 1998 survey responses showed that while chlorine was still the most widely used disinfectant, chloramine use increased from 20% in 1989 to 29.4% in 1998. In addition, 50 percent of the respondents indicated that they were considering changing their disinfection practices. The most frequent reason stated for contemplating change was to meet D/DBP regulations. The most frequent change under consideration was use of alternative disinfectants.

Currently Stage 1 of the D/DBP rule sets the MCL for TTHMs at 80 μg/L and for HAA5 at 60 μg/L based on a systemwide running annual average. In the proposed Stage 2, MCLs will remain at 80 μg/L for TTHMs and at 60 μg/L for HAA5, but utilities must meet these targets calculated as locational running averages.

CHLORAMINATION CHEMISTRY

Inorganic chloramines are produced for water treatment disinfection in reactions between ammonia (dosed and/or naturally occurring) and hypochlorous acid. When chlorine is added to water, organic chloramines are also produced in reactions between organic nitrogen compounds and chlorine. Formation of organic chloramines is undesirable because they have no biocidal potency (Feng 1966). Nitrogen organics also reduce the formation of inorganic chloramines by binding to chlorine or by the transfer of chlorine from inorganic chloramines to nitrogen organics such as amino acids (Margerum, Gray, and Huffman 1978). Another problem is that organic chloramines may react as free chlorine or ammoniacal chloramines in several of the commonly used tests to monitor these residuals. Therefore, the true nature of the disinfectant residual can be misinterpreted.

Ammonia is a highly soluble weak base. In reaction with water, ammonia forms ammonium and hydroxyl ions. The deprotonation of ammonium is shown in the following acid/base reaction:

$$NH_4^+ = NH_3 + H^+ \qquad (pKa = 9.3 \text{ at } 25\ °C)$$

When elemental chlorine gas is added to water, the chlorine is rapidly hydrolyzed to form hypochlorous acid and chloride ion in the following reaction:

$$Cl_2 + H_2O = HOCl + H^+ + Cl^- \qquad (K_{eq} = 5 \times 10^4\ M^{-1} \text{ at } 25\ °C)$$

Except for conditions of very low pH or high chloride concentration, the reaction proceeds far to the right and only a very small percentage of elemental chlorine remains.

When chlorine is added to water as liquid sodium hypochlorite, the following reaction occurs:

$$NaOCl \rightarrow Na^+ + OCl^-$$

When added as hypochlorite granules, hypochlorite is formed in the following reaction:

$$Ca(OCl)_2 \rightarrow Ca^{2+} + 2OCl^-$$

Regardless of the form of in which it is added, chlorine speciation is controlled by pH, chlorine concentration and temperature according to the following acid/base reaction:

$$HOCl = H^+ + OCl^- \qquad (pK_a = 7.5 \text{ at } 25\ °C)$$

Inorganic chloramines are formed when free chlorine reacts with ammonia in the following series of reversible stepwise reactions (Morris and Isaac1983):

$$HOCl + NH_3 = NH_2Cl + H_20 \qquad \text{(monochloramine)}$$

$$HOCl + NH_2Cl = NHCl_2 + H_2O \qquad \text{(dichloramine)}$$

$$NH_2Cl + NH_3 = NCl_3 + H_20 \qquad \text{(trichloramine)}$$

The relative amounts of the three chloramine species initially formed is dependent on pH, the Cl_2:N ratio, and temperature. Monochloramine is the desired secondary disinfectant species because of its greater stability and biocidal potency and because dichloramine and trichloramine have much lower odor thresholds. The rate of formation of monochloramine in the above equation is first order with respect to hypochlorite and ammonia and is highly pH dependent. The rate expression was reported by Morris and Isaac (1983) as follows:

$$r\ (mol / L\text{-}s) = 6.6 \times 10^8\ e^{(-1510/T)}\ [HOCl]\ [NH_3]$$

From the above equilibrium equation, it can be seen that the rate of monochloramine formation is maximized where the product of HOCl and NH_3 is greatest. The maximum NH_3 and HOCl product occurs at pH 8.4, midway between the respective pKa values of 9.3 and 7.5. At pH 8.4, the formation of monochloramine is complete in seconds to 1 minute. At higher or lower pHs the reaction is slower. Between pH 7 and 9, and at Cl_2:N weight ratios less than 5:1 (1:1 molar ratio), monochloramine is the predominant species.

As the Cl_2:N ratio increases or pH decreases, dichloramine formation occurs either in reactions between HOCl and NH_2Cl above or in the following pH dependent acid catalysis reaction in which ammonia is released:

$$2\ NH_2Cl + H^+ = NHCl_2 + NH_3 + H^+$$

The subsequent decomposition of dichloramine is primarily responsible for loss of residual. The rate of the acid catalysis reaction is increased with increasing proton donors and as pH decreases. It was reported by Valentine and Jafvert (1988) that carbonic acid and bicarbonate may catalyze to this reaction and therefore the decomposition rate of monchloramine and dichloramine may increase as alkalinity increases. Dichloramine formation is maximized in the pH range of 4.4 to 6 and at Cl_2:N ratios of 5:1 to 7.6:1. Under normal water treatment pH conditions, dichloramine formation is several orders of magnitude slower than that for monochloramine.

Breakpoint chlorination occurs when all of the ammonia present has been oxidized by chlorine. The theoretical breakpoint Cl_2:N weight ratio of 7.6:1, or molar ratio of 1.5:1, comes from the following reaction:

$$2\ NH_3 + 3\ HOCl = N_2 + 3\ H^+ + 3\ Cl^- + 3\ H_2O$$

Addition of chlorine beyond breakpoint results in a stable free chlorine residual. In practice, the Cl_2:N ratio required for breakpoint chlorination is typically higher than 7.6:1 due to other chlorine demands. It is thought that trace levels of trichloramines can be present at Cl_2:N ratios beyond breakpoint chlorination.

MICROBIOLOGICAL WATER QUALITY

One of the most important functions of a disinfectant residual is the protection of biological water quality in the distribution system. The Surface Water Treatment Rule states that all systems must maintain a minimum disinfectant residual of 0.2 mg/L leaving the plant and a detectable residual in at least 95 percent of measurements throughout the distribution system. The Total Coliform Rule requires testing for total coliforms with subsequent testing for *E.coli* for positive samples. A system is in violation if more than 5 percent of samples are positive for total coliforms or if fecal or *E.coli* is detected at a site with a second positive total coliform.

Inactivation of microorganisms and prevention of regrowth in distribution systems are affected by several factors including attachment of microorganisms to pipe surfaces to form a biofilm, penetration of the disinfectant into the biofilm, biodegradable carbon concentration, and corrosion of pipe surfaces. Recent research has suggested that chloramines are at least as effective as free chlorine in protecting microbiological water quality in the distribution system.

Several researchers have concluded that free chlorine can be ineffective in controlling bacteria in distribution systems (Tracy, Camarena, and Wing 1966, Reilly and Kippin 1983, Wierenga 1985, LeChevallier, Babcock, and Lee 1987). Conversely, several studies have reported improvements in coliform and HPC bacteria concentrations after utilities switch from free chlorine to chloramine secondary disinfection. When Huron, South Dakota switched from free chlorine to chloramines, HPC counts decreased 52 percent in the first month (Norman, Harms, and Looyega 1980).

Norton and LeChevallier (1997) monitored the water quality from two systems for one year after the systems switched from chlorination to chloramination postdisinfection. The two systems, Indiana American Water Company and Virginia American Water Company both switched to chloramines in 1993 because of problems controlling coliform bacteria.

In the Indiana American Water Company system, average HPCs for the two years before conversion were 412 cfu/mL in 1991 and 428 cfu/mL in 1992. After conversion, the yearly average HPC decreased to 31 cfu/mL. In the two years prior to conversion, 0.60 percent of 4046 samples were positive for total coliforms. After conversion to chloramines, 0.16 percent of 1868 samples were positive, almost a fourfold decrease. An increase in residual concentrations at ends of the system was also observed. In 1991 and 1992, 22 to 24 percent of the sample sites had average free chlorine residuals less than 1.0 mg/L as Cl_2. In 1993, 5.5 percent of the sites had average total chlorine residuals less than 1.0 mg/L as Cl_2.

Similar improvements in microbiological water quality were observed at the Virginia American Company. Using the less sensitive SPC agar (R2A agar was used in the Indiana measurements), HPCs averaged 7.4 cfu/mL from 1991 to 1992 and 2.6 cfu/mL in 1993 after conversion to chloramines. Prior to conversion, 22 percent of the samples were positive for coliforms. With chloramines, only 2 samples were positive for coliforms and these were both collected in the first week after conversion.

Volk and LeChevallier (2000) analyzed water quality data from 91 systems and an association between coliform counts and temperature, AOC, and type and concentration of residual. Coliform occurrences were noted when the following parameter thresholds were exceeded simultaneously: temperature > 15 °C, AOC > 100 µg/L and residuals below 0.5 mg/L free chlorine or 1.0 mg/L chloramines.

LeChevallier and colleagues (1996) performed an 18 month study of 31 systems for factors that limit microbial growth in distribution systems. Coliforms were detected in 0.97 percent of samples with free chlorine and 0.51 percent of samples with chloramine residuals.

DiGiano and coworkers (2000) performed a 12 month study to compare bacterial regrowth in a chloraminated system to that in a chlorinated system in an adjacent city. The systems made a good pair for comparison because their source water quality and distribution system temperature conditions were similar. Based on HPC data, the researchers concluded that the chloraminated and chlorinated systems provided equivalent control of bacterial regrowth. In each system, high HPC bacteria concentrations were associated with low disinfectant residual, and there was a strong correlation between low disinfectant residual and long residence time or water age.

DISINFECTION BYPRODUCT FORMATION

In research sponsored by AwwaRF, Symons et al. (1998) conducted studies on three source waters to identify the factors that affect disinfectant byproduct (DBP) formation during chloramination. Specific research goals were as follows:

- Evaluate the chemical and operating factors that influence DBP formation.
- Assess the known and unknown DBPs that are formed.
- Identify treatment steps that utilities can implement to reduce DBP formation.

The interrelationships of total chlorine residual, chloramine speciation, Cl_2:N ratio, pH, bromide concentration, and variable mixing conditions were examined for their effect of the formation of 12 monitored DBPs and total dissolved organic halogens (DOX). The 12 DBPs were the four THMs (chloroform, bromodichloromethane, dibromomochloromethane, and bromoform), HAA6 (mono-, di-, and tri- haloacetic acid, mono- and di- bromoacetic acid, and bromochloracetic acid) and the cyanogen halogens (cyanogen chloride and cyanogen bromide).

The study included source waters from the three primary utility participants, City of Austin, City of Houston, and the Metropolitan Water District of Southern California (MWDSC), and from an additional five utilities selected to the range of geographical location, water quality, and operational parameters. The project was divided into four tasks. In Task 1A, bench-scale experiments were conducted on the three primary water sources to determine effects of variable water chemistry conditions on DBP formation. Combinations of the following water chemistry conditions tested:

- pH 6, 8, and 10,
- total chlorine residual after 48 hours: 1, 2, and 4 mg/L, and
- Cl_2:N ratios of 3/1, 5/1, and 7/1.

In Task 1B, variable mixing energies and sequential chlorine addition were tested as follows:

- Low (G = 60 sec^{-1}), medium (G = 500 sec^{-1}), and high (G = 1000 sec^{-1}) mixing energies with simultaneous addition of chlorine and ammonia, and
- Chlorine addition followed by ammonia addition after a 30-second delay at low and medium mixing energies.

In Task 2, pilot studies were conducted with the three primary source waters to verify the findings off the bench-scale tests. In Task 3, five additional source waters were selected for evaluation to cover a variety of geographical locations, source water qualities, and operational variables. Full-scale operational data and selected bench-scale tests were conducted on the secondary study waters. In Task 4, analytical methods for identifying unknown DBPs were tested.

In Tasks 1A, 1B, 2, and 3, DBP residuals were measured after 48 hours incubation in simulated distribution system (SDS) tests.

The study produced the following results regarding the factors that affect DBP formation.

- There was good agreement among the bench-scale, pilot-scale, and full-scale data across the range of geographical locations indicating that the results of the study were applicable to a wide range of source water types.
- DBP formation generally increased with decreasing pH and increasing Cl_2:N ratios. Higher DBP formation at pH 6 is consistent with the theory that dichloramine, or its decomposition product, free chlorine, is the active halogenating agent.
- DBP formation was sensitive to both pH and Cl_2:N ratio. Utilities should test combinations of these two variables to formulate a chloramination strategy that balances DBP control with maintenance of adequate residual and prevention of nitrification. DBP formation was lowest at Cl_2:N ratio of 3/1, but free ammonia may be formed at ratios this low with subsequent nitrification and dissipation of residual.
- Dichloramine was present after the 48 hour SDS hold time only at pH 6, and the relative percentage of dichloramine increased as Cl_2:N ratio increased.
- Disinfectant residual was found to have little effect on DBP formation over the range of residual concentrations tested (1, 2, and 4 mg/L).

- The predominant species of HAA6 formed were the dihalogen-substituted halogens suggesting that these species were not well controlled by chloramines.
- Substantial concentrations of DOX were formed in the SDS tests under some conditions. In Task 1 B, the maximum SDS DOX concentration was 258 μg Cl^-/L in the Houston source water at pH 6 and a Cl_2:N ratio of 5/1. In Task 1B, the maximum SDS DOX concentration was 320 μg Cl^-/L in the Houston water at pH 8, a Cl_2:N ratio of 7/1, and low mixing intensity. In Task 2, the maximum SDS DOX concentration was 203 μg Cl^-/L in the Houston water at a SDS incubation pH of 6, a Cl_2:N ratio of 3/1, with prechloramination and conventional coagulation.
- The 12 specific DBPs measured accounted for a relatively small proportion of the total DOX concentration in all tests. In Task 1A, the 12 DBPs accounted for less than 35 percent of DOX on a molar basis, in Task 1B, less than 32 percent, and in Task 2, less than 45 percent. Although present and near term regulations only govern TTHMs and HAA5, utilities may elect to consider DOX formation when developing a chloramination strategy.
- Bromide addition increased the formation of bromine substituted THMs, HAA6, and CNX. The increased formation of DOX with bromide addition indicated increased degree of halogen substitution during chloramination of bromide ion containing waters. Control of DBP formation was complicated by the presence of bromide and the complexity of bromamine chemistry.
- Mixing was not as important as chemistry in controlling DBP formation. THM formation was higher at low mixing energy than at medium or high energies. THM formation was also higher for delayed ammonia addition. No correlation between DOX formation and mixing energy was observed, but DOX formation was increased by delayed ammonia addition. HAA6 and CNX formation were not affected by variation in mixing energy. It was concluded that the kinetics for THM formation might be more rapid than the kinetics for DOX formation in general.

A nonhalogenated DBP, N-nitrosodimethylamine (NDMA) has been identified as a possible formation product of chloramination. NDMA is a semi-volatile organic that has been manufactured and used commercially for decades. NDMA has been used as a soil additive to prevent nitrification, a plasticizer for rubber and polymers, and as a solvent in the fiber and plastics industry. NDMA has also been detected in a number of food products including, cheese, canned fruit, various meat products, fish and fish products, and alcoholic beverages including beer. Tobacco smoke contains high levels of NDMA. NDMA is a known carcinogen, and EPA has established a 10^{-6} cancer risk level for NDMA of 0.7 ng/L. No MCL or MCLG has been set yet by EPA for NDMA in drinking water, and the levels of exposure to NDMA as a contaminant in air and food far exceed the exposure level in drinking water.

Following discovery of NDMA in a drinking water well in 1998, California Department of Health Services (CDHS) set an action level of 2 ng/L. The action level was subsequently raised to its current level of 10 ng/L because few laboratories were capable of quantifying NDMA at the 2 ng/L level (CDHS 2002).

In bench tests simulating drinking water conditions, Eaton and Briggs (2000) found that NDMA formation was not increased by chlorine dose and was not dependent on pH. Chloramines were not tested. Najm and Trussell (2001) reported a direct correlation between chloramine dose and NDMA formation for some waters but not others. NDMA formation was not influenced by the amount of natural organic matter or organic nitrogen.

Choi and Valentine (2002) conclusively showed that NDMA can be formed by a reaction of monochloramine with dimethylamine (DMA). Dimethylamine is ubiquitous in drinking and wastewaters and was believed to be a likely NDMA precursor. A mechanism was hypothesized in which monochloramine reacts with DMA to form hydrazine that is then subsequently oxidized to form NDMA. Choi and Valentine (2002) also developed a simple kinetic model to support the hypothesis that NDMA can be slowly formed in drinking water. Formation from reactions involving other nitrogenous precursors was also suspected.

NITRIFICATION

Nitrification is a two step process in which autotrophic bacteria mediate the oxidation of ammonia to nitrate. Ammonia is oxidized to nitrite, and nitrite is oxidized to nitrate in the following sequential reactions:

$$NH_4^+ + 3/2\ O_2 = NO_2^- + H_2O + 2H^+$$

$$NO_2^- + H_2O = NO_3^- + 2H^+$$

Systems that practice chloramination are susceptible to nitrification because a small amount of ammonia is always present in a system that contains chloramines. It is estimated that two-thirds of medium and large systems that chloraminate in the U.S. experience some problem with nitrification and one –third of those systems have operational problems caused by nitrification (Wilczak et al. 1996).

The occurrence of nitrification in a distribution system can have adverse operational, regulatory, and public health effects for a utility. Since some free ammonia is typically present in systems that chloraminate, these systems are susceptible to nitrification if the loss of disinfectant residual permits the growth of ammonia oxidizing bacteria (AOB). Once nitrification begins, it accelerates the loss of disinfectant residual because the rate of chloramine decomposition increases as ammonia decreases. In the limit of very low ammonia concentrations, all residual can be lost in reactions similar to those that occur during breakpoint chlorination As chloramine is destroyed, ammonia is released providing AOB with substrate. Additional loss of chloramine may also be attributed to the formation of nitrite as a product of nitrifying organisms and further fueling the nitrification reaction.

Water quality impacts from nitrification can include loss of disinfectant residual, elevated heterotrophic bacteria counts, decrease in dissolved oxygen, and decrease in pH. Once nitrification begins, a utility is usually compelled to institute operational corrective measures such as a period of breakpoint chlorination or system flushing to stop the process. The loss of disinfection residual and increase in heterotrophic bacteria counts can affect a utility's regulatory compliance. The Surface Water Treatment Rule requires a detectable disinfectant residual at 95 percent of the sampling locations in the distribution system. The Total Coliform Rule states that no more than 5 percent of monthly samples can be positive for total coliforms. Nitrate and nitrite pose an acute health risk, particularly to infants, by inducing methemoglobinemia, in which the oxygen carrying capacity of red blood cells is compromised. Nitrate and nitrite are regulated at the entrance to the distribution system at MCLs of 10 mg/L and 1 mg/L, as nitrogen, respectively.

Kirmeyer et al. (1995) prepared an AwwaRF Report to address nitrification occurrence and control. The purpose of the study was to document the occurrence of nitrification in chloraminated systems, identify causative factors, and develop methods for controlling nitrification. Kirmeyer and coworkers concluded that nitrification is influenced by the following factors:

- Nitrification most often occurred at a pH between 7 and 9.
- Nitrification occurred at a wide range of temperatures but most often at temperatures greater than 15 °C.
- Higher chloramine residual appears to deter nitrification. Once nitrification begins however, chloramine residual disappears quickly and increasing the residual does not appear to be effective in stopping nitrification.
- Excess ammonia promotes nitrification. Ammonia concentrations were found to be 4 times higher at Cl_2:N ratio of 3/1 than at a Cl_2:N of 5/1.
- Organic compounds can exert a chloramine demand resulting in poor bacteria disinfection.
- Long distribution system residence times promote nitrification.
- Distribution system sediment and biofilms can harbor nitrifying bacteria, reduce disinfection efficiency, and enhance the likelihood of nitrification.
- Treatment plant processes that remove organic matter reduce the potential for nitrification.

The following monitoring program was recommended for detection of nitrification:

- Monitor the chemical balance of nitrogen species by measuring free and total ammonia, nitrite, and nitrate and identify changes that occur between the treatment plant and any point in the distribution system. A shift in nitrogen speciation from ammonia to nitrite and nitrate is the primary indicator of the degree of nitrification that is occurring.
- Perform heterotrophic plate counts (HPCs) with R2A agar. HPC can be an effective early warning indicator for nitrification. R2A agar was recommended over standard plate agar as being more sensitive.
- Monitor chloramine residual throughout the system regularly. A dramatic decrease in chloramine residual can indicate the onset of nitrification.
- In some utilities studied, dissolved oxygen (DO) correlated well with nitrification. DO was recommended as a secondary monitoring parameter only for utilities competent in conducting the test.

The authors rated nitrification control measures for short-term and long-term efficiency on a scale of 1 to 5, with 1 being ineffective and 5 being highly effective.

Breakpoint chlorination: Episode control – 5
Long-term control – 2

Breakpoint chlorination was rated the most effective means of controlling nitrification once an episode has started. During breakpoint chlorination, feed of ammonia, the AOB substrate, is discontinued and disinfection efficiency is increased. Many of the utilities surveyed practice breakpoint chlorination once or twice a year for a week to a month to control nitrification. Breakpoint chlorination was not considered an effective long-term solution because it provides little sustained mitigation of the factors that cause nitrification.

Increasing Cl_2:N ratio: Episode control – 3
Long-term control – 4

Reducing the ammonia by increasing the Cl_2:N ratio can be an effective control measure, but the study indicated that nitrification can take place when only small amounts of ammonia are present. This was rated a better preventive measure than as a measure for episode control. Once nitrification begins, chloramine is rapidly consumed.

Increasing residual: Episode control – 2
Long-term control – 4

Increasing the chloramine residual was rated an effective means for preventing nitrification but not for stopping nitrification once the process has started. Chloramine residuals are consumed oxidizing the nitrite before they can inactivate nitrifying bacteria. Maintaining a chloramine residual of 2 to 3 mg/L seemed to inhibit nitrifying bacteria.

Removing organics at the WTP: Episode control – 2
Long-term control – 5

Removing organics to improve distribution system biological stability was rated as a very effective method for preventing nitrification. This method has little effect once nitrification is in progress.

Distribtution system cleaning: Episode control – 3
Long-term control – 4

Cleaning the distribution system through flushing or pigging can provide effective long-term control of nitrification. Some utilities found that flushing provided only temporary relief and chloramine residuals quickly disappeared again after flushing.

Improving system residence time: Episode control – 3
Long-term control – 4

The authors concluded that decreasing system residence time can be an effective method for episode and long-term control when used in conjunction with other control measures such as breakpoint chlorination. Methods for reducing distribution system residence time include increasing reservoir turnover, installing recirculation and rechlorination facilities on storage tanks, designing new reservoirs with inlet and outlet pipes and preventing short-circuiting, looping dead-end mains, and implementing regular flushing in problem areas.

Lieu, Wolfe, and Means (1993) conducted a study to examine survival and regrowth of nitrifying bacteria after exposure to three chloramine dosages (1.7, 2.0 and 2.5 mg/L), three Cl_2:N ratios (3/1, 4/1, and 5/1), three temperatures (10, 15, and 25 °C) and two contact times (2 and 8 days). Chloramine was neutralized after the contact time except for a non-neutralized set in which chloramine residual was allowed to decay naturally. The authors concluded that the most important factor affecting bacteria regrowth was chloramine dose but that the effect of chloramine dose was influenced by temperature, Cl_2:N ratio, and contact time.

At 10 °C, after 2 days of disinfectant contact time, AOB were reduced by 1.5 log at doses of 1.7 and 2 mg/L and to nondetect level at 2.5 mg/L at all Cl_2:N ratios. After 6 weeks of incubation time, no nitrite was detected in any samples and < 10 MPN AOB were detected. It was concluded that microbiocidal efficiency was low at 10 °C, but AOB regrowth was slow.

At 15 °C, contact time and chloramine dose controlled regrowth but Cl_2:N ratio had no effect. At the 1.7 and 2 mg/L doses, nitrite was detected at 2 weeks for the 2 day exposure samples and at 3 to 5 weeks for the 8 day exposure samples. At a 2.5 mg/L dose, nitrite was detected at 4 weeks for the 2 day exposure set and was not detected for the samples exposed to chloramines for 8 days. No nitrite was detected in any of the non-neutralized samples.

At 25 °C, chloramine dose, contact time, and Cl_2:N all affected regrowth. At the 1.7 mg/L dose and 2 day exposure, nitrite appeared after 1 week for all Cl_2:N ratios. For the 8 day exposure set however, nitrite appeared after 1 week for the 3:1 Cl_2:N ratio but not until 2 weeks for 4:1 and 5:1 ratios. At the 2 mg/L dose, nitrite was not detected in the 8 day exposure or in the non-neutralized set at 4:1 and 5:1 Cl_2:N ratios. At the 2.5 mg/L dose, nitrite was detected only in the 2 day exposure sample with the 3:1 Cl_2:N ratio.

Disinfection efficiencies were similar at 10 and 15 °C, but regrowth occurred in twice as many samples at 15 °C. It was concluded that nitrification potential was higher at 15 °C. The biocidal effectiveness of chloramines at 25 °C increased when Cl_2:N ratio increased from 3:1 to 4:1 and 5:1. Consequently, there were fewer samples with regrowth at 25 °C than at 15 °C. It was concluded that increasing the Cl_2:N ratio could provide nitrification control for some situations. Regrowth was highest in the 25 °C tests confirming the expected optimal temperature range for regrowth of 25 to 30 °C. Finally it was concluded that when nitrite levels are first detectable, a removal of greater than 2 log is necessary to prevent regrowth.

McGuire, Lieu, and Pearthree (1999) conducted field and lab tests using chlorite ion to control nitrifying bacteria. In the lab tests, chlorite dosages as low as 0.05 mg/L achieved 3 to 4 log inactivation of AOB over several hours. In field studies of systems using chlorine dioxide for primary disinfection, systems with chlorite showed lower concentrations of nitrifying bacteria than systems with chlorite. The authors suggested that chlorite could be added to the plant effluent in low enough concentration to avoid exceeding the 0.8 mg/L MCL for chlorite.

Wolfe et al. (1990) conducted a study to determine the factors that affect the occurrence of AOB in chloraminated distribution systems. The authors concluded that the occurrence of AOB was seasonally dependent with higher concentrations in the summer months. AOB correlated highly with temperature and HPC bacteria. No AOB were found in the reservoir when temperatures were below 16 to 18 °C. AOB were found throughout the distribution system and were found in higher concentrations in sediment than in pipe wall biofilms. AOB were 13 times more resistant to chloramines than chlorine.

Regan, Harrington, and Noguera (2002) also found an increase in HPC bacteria at the onset of nitrification and suggested HPC measurement with R2A agar as an early warning indicator.

CONVERSION

Over the last 20 years, several utilities have converted from free chlorine to chloramines to reduce THM formation in their systems. The Metropolitan Water District of Southern California (MWDSC) converted to chloramines in 1984 and documented their experience in several publications. These publications provide a clear example of the steps conducted by one large utility in converting to chloramines as well as presenting some of the problems that arose and their solutions.

Prior to conversion, MWDSC conducted a study to evaluate the feasibility of switching to chloramination (Kreft et al. 1985). MWDSC first tried other steps to reduce THM formation. Preoxidation with potassium permanganate was not successful. Enhanced coagulation did not provide sufficient THM reduction. Ozone was deemed too expensive and chlorine dioxide was discounted because of concerns over chlorite and chlorate formation. Pilot testing with chloramines was successful.

MWDSC conducted a feasibility study to evaluate the following factors:

- Chemistry and DBP formation.
- Disinfection capacity.
- Taste and odors.
- Bacterial growth in uncovered reservoirs and effects on swimming pools.
- Maintenance and effectiveness of disinfectant residual.
- Health effects of chloramines.

- Impact on member agencies that use chlorination, i.e. effect of blending.
- Technological feasibility, how to implement.
- Costs.

MWDSC used jar testing to develop a chlorine breakpoint curve for their water to determine the range of chlorine to ammonia feed ratio necessary to produce monochloramine. Monochloramine residual increased to a Cl_2:N ratio of about 5:1, and reached a minimum at a ratio of 10.5:1. Free chlorine residual began to appear at a ratio of 11:1. MWDSC concluded that the Cl_2:N ratio should be maintained at less than 5:1. Some dichloramine was formed prior to breakpoint and after breakpoint the trichloramine ratio increased.

MWDSC concluded that chloramines would provide adequate disinfection capacity. MWDSC noted that at their operating pH of 8.1 to 8.2, OCl^- was the predominant chlorine species after chlorination. It was noted that the difference in disinfectant potency reported between monochloramine and hypochlorite is less than that between monochloramine and hypochlorous acid. Additionally MWDSC pointed to work by Collins and Selleck (1972) and Kruse, Snead, and Olivieri (1981) that indicated that total chlorine is as effective as free chlorine in inactivating bacteria at long contact times. MWDSC also relied on successful secondary disinfection reported at over 70 utilities practicing chloramination.

To assess taste and odor effects of conversion to chloramination, MWDSC conducted a side-by-side flavor profile analysis (FPA) of chlorinated and chloraminated water. The FPA panel concluded that the chloraminated water had a much less offensive odor.
MWDSC conducted educational seminars with member agencies, hospitals, patients and dialysis patients to explain the potential adverse effects of chloramines on dialysis patients. Information was distributed to pet stores, zoos, aquariums and the public about protecting fish from chloramines.

Many of its member agencies bought water from MWDSC to augment their own chlorinated supplies. MWDSC met with member agencies to discuss blending issues, particularly the avoidance of loss of disinfection residual due to breakpoint chlorination. MWDSC developed a computer model to predict the effects of blending (Barrett, Davis, and McGuire 1985). Jar testing was used to calibrate the model. The model was then used to predict acceptable blends of free chlorine and chloramines for each of its member agencies to prevent breakpoint chloramination and odors caused by dichloramine ad trichloramine.

In their cost analysis, MWDSC estimated that conversion to chloramines would cost much less than GAC treatment, an alternative method of reducing THMs by removing precursors.

Following conversion in 1984, MWDSC staff summarized the practical aspects of implementing chloramination (Dennis, Rauscher, and Foust 1991) MWDSC selected aqueous ammonia as used by Denver and Philadelphia because it is less expensive to store, less hazardous to transport, and permits use of simple metering pumps. MWDSC initially chose concurrent addition of chlorine and ammonia to limit THM formation and because disinfection CT was obtained with the three hour contact time.

MWDSC conducted an extensive planning program prior to actual conversion. An aqueous ammonia manual was prepared that included feed equipment, piping layouts, analysis and safety and maintenance issues. A formal request was sent to the state of California to make the change. A series of 6 workshops was conducted with the California Department of Health Services (CDHS), member agencies, and local health departments.

An extensive notification program was conducted that coordinated notification of CDHS, 5 county health departments, the Federally funded End-stage Renal Dialysis Coordinating Council, local nephrologists, and all member agencies. Local pet shops and koi clubs were also notified. Over 60 newspaper articles, press releases, and radio and television announcements were made, and a telephone hotline was set up to answer questions.

Two task groups were formed to facilitate implementation. One was created to procure equipment, monitor construction schedule, test pumping equipment, and ensure the readiness of the WTPs and reservoir facilities to feed ammonia. The other task group was responsible for training operators and lab personnel. Ten training seminars were conducted.

Potential problems were uncovered and remedied in the pump-testing program. The Buna-N rubber stators failed within a week and were replaced with ethylene propylene diene monomer material. Leakage from the packing box on the pump was detected. The packing boxes were replaced with mechanical seals.

Immediately after conversion, some dialysis centers reported problems removing chloramines and a few cases of hemolytic anemia were reported. MWDSC immediately switched back to chlorination. In cooperation with CDHS and county health departments, interviews were conducted with dialysis centers and all dialysis centers were inspected. Standards for chloramine removal were developed and a plan was formulated for each facility. After changes had been made, inspectors went back to each facility and conducted another inspection. This process took approximately 6 months at which time chloramination was resumed.

In June of 1985, MWDSC began receiving complaints of fishy odors. Jar tests indicated that the odor problem could be eliminated with 1 hour of free chlorine contact time. Prechlorination was instituted with chlorine added 1 hour ahead of ammonia.

Ten weeks after conversion, in 1984, MWDSC observed nitrification in the reservoirs. In 1987 and 1988, nitrification occurred in the distribution system. MWDSC developed a nitrification control program that included early warning sampling, decreasing distribution system residence time, increasing the Cl_2:N ratio from 3:1 to 5:1, and maintaining a free chlorine residual for 30 days each year.

PROBLEMS

Conversion to chloramination can help utilities reduce THM formation and increase residual longevity, but there are also potential problems associated with chloramination that utilities must avoid. Chloramines pose a risk to dialysis patients in the form of methemoglobinemia and acute hemolytic anemia. Chloramines can cause methemoglobinemia in dialysis patients by oxidizing Fe^{2+} to Fe^{3+}, methemoglobin, in the blood (Klein 1986). Methemoglobin is incapable of carrying oxygen or carbon dioxide. Chloramines can also denature proteins within the red blood cell forming aggregates (Heinz bodies).

In 1981, there was a reported outbreak of Heinz-body positive hemolytic anemia in a Sydney, Australia hospital (Caterson et al. 1982). It was found that the activated carbon filter system was inadequate. After increasing the filter length, the problems disappeared. Kjellstrand and coworkers (1974) found that excess levels of chloramines led to methemoglobin formation, Heinz-body formation, and hemolysis in dialysis patients. It was found that chloramines at a concentration of 0.25 mg/L invariably caused hemolysis in dialysis patients.

When a utility converts to chloramines, aquarium owners must take precautions to protect their fish. Chloramines are introduced directly to the bloodstream of fish through their gills and pose some of the same health risks incurred by dialysis patients, including hemolysis. Aquarium owners must protect their fish against exposure to chloramines and the buildup of ammonia. In preparation for conversion to chloramination, utilities typically take steps to educate the public and pet storeowners regarding methods to protect fish.

Chloramines have also been found to degrade elastomers more aggressively than chlorine. Reiber (1993) tested elastomer coupons in an accelerated life cycle protocol with chlorine and chloramines. Reiber reported the following:

- Chloramines were more aggressive in degrading elastomers than equivalent concentrations of chlorine, and dichloramine was more aggressive than monochloramine. It was theorized that chloramines achieved greater penetration into the elastomer material due to less demand than exerted for chlorine at the elastomer surface.
- Excess ammonia addition or the presence of free ammonia was not responsible for degradation. Coupons exposed to solutions of ammonia at different pH values showed the same amount of degradation as dechlorinated tap water.
- Forms of elastomers varied widely in their resistance to degradation. The materials most susceptible to attack were the natural isoprene and synthetic isoprene derivatives. Synthetic polymers engineered for chemical resistance performed well in the life cycle tests.
- A water swell test was found a reliable indicator for material degradation. The tendency of elastomer material to absorb water was found to correlate strongly to breakdown of material strength properties such as maximum stress and strain.
- Elastomer degradation was found to be sensitive to temperature and to occur more rapidly with increased temperature.

APPENDIX B
AwwaRF CHLORAMINATION SURVEY FORM

UTILITY ________________________________
Address ________________________
____
____

Name of Person filling out this form: ________________________
Phone no.: __________________

1. **Does your system use chloramines for a residual within the distribution system?**
   yes ___ **no** ___ *(if you answered no, do not complete this form, just return it)*

2. **If you use chloramines:**
   a. **Why do you use chloramines?**
      To reduce DBP levels? yes ____ no ____
      When using free chlorine, our THM levels ranged from _____ mg/L to ____mg/L
      With chloramines, our THM levels now range from _____ mg/L to _____ mg/L

      To improve the taste of the water? yes ____ no ____
      To help control biofilms in system? yes ____ no ____
      Other yes ____ no ____ *(Please comment)*

   b. **How long have you used chloramines?** _______ **years**

3. **Our source of water is primarily surface water. yes ____ no ____**

4. **We use both surface water and groundwater. yes ___ no ___**

5. **We have _____water treatment plants in our system.**
[Please list the capacity of each treatment plant in millions of gallons per day (mgd)]

6. **Did you experience any significant start-up problems at either the treatment facility or within the distribution system when you began using chloramines? yes ___ no ___**

   If yes, please briefly describe them:

7. Do you sometimes use short periods of free chlorine instead of chloramines for disinfection within the distribution system? yes ___ no ___

If yes,

How often? ______________________

How long? ______________________

What free chlorine concentration is used ? ______ mg/L

Why do you switch from chloramines to free chlorine?

Are industrial and medical customers notified of this change? yes ___ no ___

If yes, how?

Is the public notified of this switchover? yes ___ no ___

If yes, how?

Do you flush the system during the switchover? yes ___ no ___

Do you do additional sampling during this period? yes ___ no ___

If yes, what parameters are measured?

Do you ever measure DBPs during this period of free chlorine use? yes ___ no ___

Is your State Agency notified of the switch to free chlorine? yes ___ no ___

If yes, what is their policy regarding DBP sampling when free chlorine is used?

Are customer complaints or reactions monitored during the duration of the free chlorine use? yes ___ no ___

How much preparation time is involved in switching to free chlorine?

________ man-hours

How many people are involved in preparing for the switch to free chlorine? ____

If you have an estimated additional cost of this program, please indicate below.

$ __________

Do you have any documented benefits of this switchover program? yes ____ no ___

If yes, the benefits are:

8. **The treatment process we use at our largest plant has the following unit processes:**
(please check only those unit processes that apply)

aeration ___
presedimentation ___
coagulation ___
flocculation ___
sedimentation ___
granular media filtration ___
membrane filtration ___
disinfection ___
with free chlorine gas ___
with free chlorine liquid ___
with combined chlorine ___
with chlorine dioxide ___
with ozone ___
with UV ___
lime softening ___
recarbonation ___
other processes used:

Chemicals normally used:
anhydrous ammonia ___
aqua ammonia ___
ammonium sulfate ___
alum ___
polymer ___
ferric salts ___
ferrous salts ___
soda ash ___
sodium hydroxide ___
sulfuric acid ___
fluoride ___
powdered activated carbon ___
potassium permanganate ___
corrosion inhibitor ___
other chemicals:

Are there any problems or concerns with the storage, feeding or mixing of your chlorine or ammonia chemicals? yes ___ no ___ If yes, please describe.

9. **Where you have available data, please fill in typical water quality data for your treated water as it enters the distribution system.**

Constituent	**Winter**	**Summer**
Water temperature, °C		
pH, units		
Alkalinity, mg/L		
Total hardness, mg/L		
Total dissolved solids, mg/L		
Nitrate, mg/L as N		
Free ammonia, mg/L as N		
Total organic carbon, mg/L		
Free chlorine residual, mg/L		
Total chlorine residual, mg/L		

10. We target our chlorine:ammonia ratio at ___:____.

11. Do you have difficulty maintaining your target chlorine to ammonia ratio?
yes ___ no ___

12. Residence (travel) time within our distribution system averages ______ days and is a maximum of about _______ days before it reaches the end of our distribution system.

13. Some concerns (mark with letter "C") and problems (mark with letter "P") we have about using chloramines are:
(please mark all those that apply using letters **C** and **P**)

nitrification ___ uses more chlorine ___ disinfection efficiency ___
blending ____ taste and/or odors ___ using ammonia ___
customer acceptance ___ operations ___ need to flush pipes ___
maintaining chorine residuals in distribution system ___ biofilms ___
increase in microorganisms ___ free ammonia at some locations ___
corrosion ___ degradation of distribution system materials ___
medical issues ___ aquariums ___ fish ponds ___
blending with wholesale customer systems ___
other:

14. Do you apply free chlorine before converting to chloramines? yes ___ no ___
If yes, chlorine is added at ___________________ with approximately _______ minutes hydraulic residence time (detention time) before ammonia is added at ________________.

15. The pH of the water where we add ammonia is normally about _________.

16. Do you blend chlorinated water with chloraminated water? yes ___ no ___

17. Chloramine residuals at the far ends of the distribution system are about ____ mg/L.

18. Do you have any current difficulties with using chloramines? yes ___ no ___

If yes, please briefly describe them:

19. Do you practice booster chloramination or rechlorination in the distribution system?
yes ___ no ___ If yes, how is this controlled?

20. Do you provide information regarding your chloramine residual to new customers or industries? yes ___ no ___ If yes, what information do you provide?

21. Do you make any seasonal changes in your water quality monitoring program with regard to chloramines? yes ___ no ___ If yes, please describe.

22. Please give us any other comments you would like to make about using chloramines.

APPENDIX C
SURVEY PARTICPANTS

Abilene, TX
Alameda Co. Water District, Fremont, CA
Ann Arbor, MI
Appromattox River Water Authority, Petersburg, VA
Aurora, CO
Austin, TX
Beaumont, TX
Board of Public Utilities, Kansas City, KS
Brownsville, TX
B-Y Water District, Tabor, SD
Charleston Commissioners of Public Works, Hanahan, SC
Chesterfield Co. Utilities, Midlothian, VA
City Water, Light & Power, Springfield, IL
Concord, NH
Consumers PA Water Co., Sharon, PA
Dallas, TX
Elgin, IL
Escondido-Vista, Escondido, CA
Fort Wayne, IN
Fort Worth, TX
Gulf Coast Water Authority, Texas City, TX
Hammond, IN
Hollywood, FL
Houston, TX
Independence, MO
Indiana-American, Gary, IN
Indiana-American, Muncie, IN
Indiana-American, Newburgh, IN
Jackson, MS
Jefferson Parish, Jefferson, LA
Lawton, OK
Lincoln, NE
Louisville, KY
Manatee Co., Bradenton, FL
Melbourne, FL
Metropolitan Water District of Southern California, Los Angeles, CA
Miami-Dade Water and Sewer, Miami, FL
Milwaukee, WI
Minneapolis, MN
Municipal Authority of Westmoreland Co., Saltsburg, PA
North Miami Beach, FL
North Texas Municipal Water District, Wylie, TX
Odessa, TX

Palm Beach Co., West Palm Beach, FL
Philadelphia Suburban Water Co., Bryn Mawr, PA
Portland, ME
Portland, OR
Pueblo, CO
Richmond, VA
Santa Clara, CA
San Diego, CA
Sioux Falls, SD
St. Louis, MO
St. Paul, MN
Sweetwater Authority, Spring Valley, CA
Tampa, FL
Topeka, KS
Waco, TX
Water District No. 1 of Johnson Co. Kansas, Kansas City, KS
Wichita Falls, TX
Wichita, KS
York, PA
Zone 7 Water Agency, Livermore, CA

APPENDIX D
CHEMISTRY OF CHLORAMINES

Chloramines have long been used to provide a disinfecting residual in distribution systems where it is difficult to maintain a free chlorine residual or where disinfection by-product (DBP) formation is of concern. While chloramines are generally considered less reactive than free chlorine, they are inherently unstable even in the absence of reactive substances. These reactions, often referred to as "auto-decomposition", always occur and hence define the maximum stability of monochloramine in water. The effect of additional reactive material must be measured relative to this basic continual loss process..

The net reaction, simplified from the approximately 14 individual reactions that govern it, can be written as,

$$3\ NH_2Cl \rightarrow N_2 + NH_3 + 3\ Cl^- + 3\ H^+$$

In essence, auto-oxidation is nothing more than the "break point" reaction, occurring however, at a much reduced rate. While relatively slow, it may still account for a substantial loss of residual in distribution systems characterized by long residence times.

A thorough understanding of chloramine formation and auto-decomposition reactions is important because this system is highly dynamic and especially susceptible to perturbations in pH, ammonia, and chlorine concentrations. The complex set of reactions describing this govern the initial formation of various forms of chloramine species and their transformation to other forms with time, loss of chloramines by auto oxidation, and demand reactions as well as formation of DBPs.

A kinetic model describing chloramine formation and auto-decomposition was developed (Jafvert and Valentine, 1992; Vikesland et. al. 1998). This model (Table D.1) is based on studies of isolated individual reactions and on observations of the reactive ammonia-chlorine system as a whole. The model includes three principle reaction schemes as well as pertinent equilibrium reactions:

1. Speciation reactions of HOCl with ammonia and the chlorinated derivatives of ammonia, and the corresponding hydrolysis reactions.
2. Disproportionate reactions of chloramine species and the corresponding back reactions.
3. Redox reactions that occur in the absence of measurable free chlorine.

Chloramine decay can be adequately predicted by the detailed mechanistic model. However, its use is relatively complicated. Chloramine decay in the absence of demand reactions, may, under most conditions, be estimated by a simple second order relationship. The integrated form can be given in terms of a single coefficient (k_{VCSC}) as :

$$\frac{1}{[NH_2Cl]} - \frac{1}{[NH_2Cl]_0} = k_{VCSC}\, t$$

The value of k_{VCSC}(Valentine Chloramine Stability Coefficient) increases with decreasing pH and initial chloramine concentration that determines total free ammonia in the system. It also increases with increasing total inorganic carbon and temperature. It can be rationalized in terms of measurable parameters and known rate constants and therefore does not need to be measured for each water.

Table D.1

Modeling chloramine decomposition kinetics

	Reaction	Rate Constant	Reference
(1)	$HOCl + NH_3 \Longrightarrow NH_2Cl + H_2O$	$k_1 = 1.5x10^{10}\ M^{-1}\ h^{-1}$	Morris and Isaac (1983)
(2)	$NH_2Cl + H_2O \Longrightarrow HOCl + NH_3$	$k_2 = 7.6x10^{-2}\ h^{-1}$	Morris and Isaac (1983)
(3)	$HOCl + NH_2Cl \Longrightarrow NHCl_2 + H_2O$	$k_3 = 1.0x10^{6}\ M^{-1}\ h^{-1}$	Margerum et al. (1978)
(4)	$NHCl_2 + H_2O \Longrightarrow HOCl + NH_2Cl$	$k_4 = 2.3x10^{-3}\ h^{-1}$	Margerum et al. (1978)
(5)	$NH_2Cl + NH_2Cl \Longrightarrow NHCl_2 + NH_3$	k_d^{*}	Jafvert (1985)
(6)	$NHCl_2 + NH_3 \Longrightarrow NH_2Cl + NH_2Cl$	$k_6 = 2.2x10^{8}\ M^{-2}\ h^{-1}$	Hand and Margerum (1983)
(7)	$NHCl_2 + H_2O \Longrightarrow I$	$k_7 = 4.0x10^{5}\ M^{-1}\ h^{-1}$	Jafvert and Valentine (1987)
(8)	$I + NHCl_2 \Longrightarrow HOCl$ + products	$k_8 = 1.0x10^{8}\ M^{-1}\ h^{-1}$	Leao (1981)
(9)	$I + NH_2Cl \Longrightarrow$ products	$k_9 = 3.0x10^{7}\ M^{-1}\ h^{-1}$	Leao (1981)
(10)	$NH_2Cl + NHCl_2 \Longrightarrow$ products	$k_{10} = 55.0\ M^{-1}\ h^{-1}$	Leao (1981)
(11)	$HOCl \Longleftrightarrow H^+ + OCl^-$	pKa= 7.5	Snoeyink and Jenkins (1980)
(12)	$NH_4^+ \Longleftrightarrow NH_3 + H^+$	pKa= 9.3	Snoeyink and Jenkins (1980)
(13)	$H_2CO_3 \Longleftrightarrow HCO_3^- + H^+$	pKa= 6.3	Snoeyink and Jenkins (1980)
(14)	$HCO_3^- \Longleftrightarrow CO_3^{-2} + H^+$	pKa= 10.3	Snoeyink and Jenkins (1980)

$$* k_d = k_H[H^+] + k_{H_2CO_3}[H_2CO_3] + k_{HCO_3}[HCO_3^-]$$

where $k_{H_2CO_3} = 4x10^4\ M^{-2}h^{-1}$, $k_{HCO_3} = 800\ M^{-2}h^{-1}$

The Valentine Chloramine Stability Coefficient can be calculated for any water source to define the limit the chloramine stability. Utilities may use this to calculate the effects of water quality and chloramination practices (e.g. Cl/N ratio) on chloramine stability. Observed rates that exceed those predicted k_{VCSC} could also be used to point to problems in the system such as the existence of oxidizable iron or organics. Observed rates of loss that are less than this may point to the existence of relatively stable organic chloramines. Some values of k_{VCSC} for various water quality conditions are given in Table D.2.

Table D.2

Valentine Chloramine Stability Coefficient, k_{VCSC}, for various water quality conditions

Monochloramine mg/L as Cl_2	Cl2:NH_3-N mg/L:mg/L	pH units	k_{VCSC} L/mg·d
4	3	6.5	0.0843
4	3	7	0.0284
4	3	7.5	0.0094
4	3	8	0.00307
4	3	8.5	0.00105
2	3	6.5	0.15
2	3	7	0.0494
2	3	7.5	0.016
2	3	8	0.00521
2	3	8.5	0.00176
4	1.5	6.5	0.0374
4	1.5	7	0.0136
4	1.5	7.5	0.00464
4	1.5	8	0.00155
4	1.5	8.5	0.000545
2	1.5	6.5	0.0565
2	1.5	7	0.0196
2	1.5	7.5	0.00656
2	1.5	8	0.00217
2	1.5	8.5	0.000751
4	5	6.5	3.55
4	5	7	1.13
4	5	7.5	0.359
4	5	8	0.115
4	5	8.5	0.0384
4	4.5	6.5	0.381
4	4.5	7	0.122
4	4.5	7.5	0.0392
4	4.5	8	0.0127
4	4.5	8.5	0.004

APPENDIX E
SITE VISITS

Thirteen water systems that use chloramines were visited as part of this project. Each visit was a personal interview conducted with a knowledgeable representative(s) of the system visited. Most of the utilities visited requested that they not be specifically identified by name. There were multiple reasons for this request, but security issues were often cited. Consequently, the majority of the systems are designated in this appendix by a letter and are not further identified. Three systems agreed to be named; they are:

- Gulf Coast Water Authority of Texas City, TX
- WaterOne of Lenexa, KS, and
- City of Wichita Falls, Wichita Falls, TX

Regions of the U.S. are represented by the 13 water systems as follows:

- 6 systems from the southwest
- 2 systems from the southeast
- 4 systems from the east
- 1 system from the mid-west

MEETING SUMMARY – Gulf Coast Water Authority, Texas City, TX

AwwaRF Project No. 2847 **B&V Project No. 131832.100**

February 12, 2003

Participants: Ms. Nancy Matthews, Gulf Coast Water Authority
Mr. Steve Carpenter, Gulf Coast Water Authority
Dr. Lee Harms, B&V – Kansas City
Mr. Gary Smith, B&V – Houston

RE: Site visit for chloramination project

Location: Gulf Coast Water WTP
3630 FM 1765
Texas City, TX

General

Gulf Coast Water Authority is somewhat unique in that they wholesale water to several customers; they do not have residential customers of their own. Consequently, most functions such as monitoring, flushing, etc. are carried out by their customers, not Gulf Coast Water. Treated water is sold to 13 communities (180,000 population) and one industrial user. An additional 35 to 40 mgd of raw water is bypassed to refineries. Surface water is transported approximately 200 miles by canals for raw water into the plant; only recently have they had any ground water to use in their system. The intent is not to use the wells except in an emergency as the water is of poor quality and has a bad odor.

Gulf Coast Water does treat to assist their customers in meeting requirements of the SDWA. For example, they are concerned about water quality issues such as low chlorine residuals and high lead values, and they initiate measures to deal with these conditions. A zinc orthophosphate is fed to help one customer meet the lead action level.

Nitrification

Nitrification does occur within their system as depicted periodically by low chlorine residuals. It is primarily one customer that experiences low chlorine residuals. When this occurs, they notify the State, and then switch to a free chlorine burn for 2 to 3 weeks at 2.5 mg/L free chlorine. This takes care of the low residual problem and then chloramines are used once again. The State is notified about the burns so DBP samples are not collected during this period. It is not clear if this procedure will be allowed under the Stage 2 DBPR.

Controlling Cl_2:NH_3 ratio

The target Cl_2:NH_3 ratio is 3:1 as residual and they achieve normally achieve this by dosing at 4:1. Their free ammonia is 0.1 to 0.15 mg/L. They caution against overfeeding ammonia as this was a problem they had initially and many of their coliform samples came back with confluent growth and could not be counted. They run the free chlorine DPD test to be sure they are feeding enough ammonia and are converting all their residual to chloramines. When they first converted to chloramines, they ran amperometric titrations to determine if all the chloramines were present as monochloramine.

Operation

The only operational problems they experience are if the shut down procedure is not followed correctly. They feed ammonia gas directly into a pressurized header. If the pump is stopped prior to isolating the feed line with the shut-off valve, the pressure from the header will back water into the transfer line. They use a straight pipe inserted into the water line and do not use a diffuser.

They currently use ammonia gas but are considering switching to liquid ammonium sulfate as a safety precaution.

Blending

They currently do not blend ground water and surface water.

Water Quality

The raw water quality is fairly poor having traveled a long distance in the canals. It cleans up well using solids contact clarifiers and ferric sulfate as a coagulant. Settled water turbidities run 1 to 2 NTU. They have used an Aromascan instrument to monitor for odors but are not using it now.

Disinfection

Disinfection is accomplished using co-disinfection with chlorine dioxide and chlorine fed at the head of the clearwell. About 60 minutes of contact is achieved within the clearwell. Ammonia is fed at the exit of the 1.1 MG clearwell for conversion to chloramines. Care is taken during analysis to purge the chlorine prior to measuring the chlorine dioxide concentration. Chlorine dioxide is the primary disinfectant used to obtain CT. Chlorine dioxide is also fed at the head of the plant to control algae and tastes and odors. Disinfection of new and repaired lines has not been a problem because of using chloramines.

MEETING SUMMARY – WaterOne, Lenexa, KS

AwwaRF Project No. 2847

B&V Project No. 131832.100
March 27, 2003
Rev. April 9, 2003

Participants: Mr. Chuck Weber, WaterOne
Mr. Dale Tapp, WaterOne (via phone)
Dr. Lee Harms, B&V

RE: Site visit for chloramination project

Location: Hansen WTP
7601 Holliday Dr.
Kansas City, KS

General

The meeting was held to discuss some operating issues associated with chloramination. Two water treatment plants comprise the Hansen treatment facility: Facility 1, the older of the two has a 55 mgd maximum capacity; and Facility 2 has a capacity of 110 mgd. Water for these two plants is primarily from the Kansas River for Facility 1 and from the Missouri River for Facility 2, although some ground water can also be added to either or both plants. The percentage of ground water treated is small compared to the surface source but it is especially valuable in the winter months as the ground water is warmer and assists in preventing freezing in the softening basins. These facilities use chlorine dioxide for disinfection and lime for softening. Filters are located indoors but softening basins are external and are not covered. Chloramines have been successfully used for almost 30 years without difficulty. A good disinfectant residual is carried through the entire system and HPC counts are exceptionally low (<10 /mL).

Nitrification

Nitrification is not a problem for Water District No. 1 of Johnson Co., KS (WaterOne) as determined by minimal drop in chlorine residual through the system, few taste and odor complaints, and low HPC counts (biofilm indicators). Nitrate does fluctuation some during the year but it is a function of nitrates in the raw water not nitrification. Ammonia and nitrate are sometimes measured at various locations within the distribution system but they are not monitored on a regular schedule. From information elsewhere, it is also possible that the high finished water pH (~ 9.5) also assists in retarding the metabolism of any nitrifying organisms that may try to grow within the system. WaterOne does not practice a free chlorine burn and the water quality evidence supports this decision.

Controlling Cl_2:NH_3 ratio

Chlorine is added prior to the filters and ammonia is added after filtration but before treated water storage. Disinfection has previously been accomplished using chlorine dioxide. Controlling chemical feeds to a specific Cl_2:NH_3 ratio is not practiced by the District. Instead, a set dose of 1.4 mg/L ammonia has historically been used with a chlorine dose of about 4 mg/L. This would compute to a Cl_2:NH_3 somewhat less than 3:1 but should ensure monochloramine residuals being formed.

Operation

Major difficulties with the chemical feed systems have not been experienced although some solids build-up has been occasionally noted at the ammonia feed location. The solids did react with dilute acid so they may have been a carbonate precipitate resulting from the pH change immediately at the injection site. Consideration is being given to using a diffuser in lieu of the current arrangement. Presently this feed system uses a small pipe which directly injects the ammonia solution (29 percent aqua ammonia) into the water. Injection is accomplished by maintaining a constant low pressure on the ammonia solution tank and the pressure difference feeds the ammonia through a Rota meter. Some changes in pressure are caused by daily environmental conditions such as sunlight directly on the tank.

Blending

Chlorinated ground water is not blended with chloraminated surface water. Both ground water and surface water pass through the treatment train.

Water Quality

No water quality issues associated with the use of chloramines are apparent.

Disinfection

As previously noted, disinfection is accomplished with chlorine dioxide, not chlorine or chloramines. Although not required to remove chlorite to meet the MCL, ferrous ion is added to keep the chlorite ion concentration at less than 0.10 mg/L at the tap.

MEETING SUMMARY – Wichita Falls, TX

AwwaRF Project No. 2847 **B&V Project No. 131832.100**

Participants:

Daniel Nix, Water Plant Superintendent, City of Wichita Falls
Dr. Mark Waer, B&V-Phoenix

RE: Chloramination Project Site Visit

Date: February 6, 2003, from 10:45 to 11:45 a.m.

Location: Wichita Falls Cypress Water Treatment Plant
Big Ed Neal Drive
Wichita Falls, TX

INTRODUCTION

This meeting was held to discuss operational issues associated with chloramination of drinking water for disinfection. I was given a tour of the chloramination facilities at the Wichita Falls Cypress Water Treatment Plant, and then met with Mr. Daniel Nix, the plant superintendent to discuss these issues.

DISINFECTION

In spring, summer, and fall, chlorine is added to the raw water main as it enters the chemical building basement. Ammonia is added after the raw water main splits to feed the old plant and new plant. Primary disinfection is achieved with chloramines through the entire treatment plants.

During winter, the contact time through the plant is insufficient to maintain primary disinfection with chloramines, so chlorine dioxide is injected upstream of the plant and a residual is measured within the plant to calculate CT.

Water temperature is measured throughout the year and in the late fall when it falls below a set value, chlorine dioxide disinfection is started up. In the spring, when the water temperature rises above that value, chloramine primary disinfection is practiced.

CHLORINE TO AMMONIA RATIO

The plant maintains a 4:1 chlorine to ammonia ratio based on chlorine and ammonia dosages. They do not account for chlorine demand, although there is probably less than one minute between chlorine and ammonia addition. The trihalomethane formation during this interval is approximately 14 μg/L. The plant has experimented with other ratios, but 4:1 appears to work best for them. During winter time when flows are low, the chlorinators and especially the ammoniators are oversized. This leads to problems with ammonia overfeed. There are projects underway to rectify the equipment problems.

NITRIFICATION

Most years the system does experience a few isolated pockets of nitrification. Years ago, nitrification was a larger problem, but they learned that they could control it by cycling the water through their storage tanks and keeping the water age in the system down. They detect nitrification mainly by chlorine residual decay in the system. They verify by pH measurement and mainly by nitrite concentrations. They have seen nitrite concentrations as high as > 2 mg/L. They also run a test for nitrifying bacteria (nitrifying BART).

This past summer (2002) was very mild, and they had trouble keeping the water age low. Nitrification was a systemic problem this year and they had to resort to control measures. The pH of the system was raised to 9.2 from its normal 8.5. After about a day of this, the nitrite concentrations dropped by about 50%. Ammonia was turned off at the plant and the system was subjected to 4 mg/L of free chlorine (leaving the plant) for about 10 days. The residual at the end of the system was about 2.5 mg/L at the end of this treatment. Nitrite levels were non-existent after about 24 hours of free chlorine. Free chlorine in the system caused numerous taste and odor complaints from the public. The use of high pH and free chlorine is being discussed as a summer standard operating procedure, either on an "as needed" basis or as a regular occurrence once each summer.

During this event, the plant got a call from a chicken restaurant in its distribution system. The restaurant was getting complaints from its customers that the chicken was pink even though it was cooked. They had contacted their "home office" and were told that there was a problem with the water and the preservative used on the chicken reacts to form a pink color in the presence of nitrite. The lab took a water sample at the restaurant and found 2.6 mg/L of nitrite.

The lab analyzes samples from the plant effluent for free ammonia, but not the distribution system. Free ammonia is normally not found during summer, but during winter, levels as high as 1.5 mg/L can be measured. This is due to the oversized feed system.

OPERATIONS ISSUES

The plant feeds all of its ammonia in the raw water, but only feeds about 75% of its chlorine there. The remainder is added to water after filtration. They have found that the free ammonia interferes with the coagulation process. Their process is a single stage softening process at pH 10.5 with ferric addition. The ammonia has been found to interfere with the ferric coagulation under these circumstances. The plant has turned off the ammonia and observed a dramatic clearing of the basins as the result. The upgrade projects at this plant will install separate ammonia feeds to the post to eliminate this problem.

WATER QUALITY

In winter time, the plant sees higher than normal ammonia concentrations in the raw water, but this doesn't appear to interfere with disinfection.

BLENDING WITH WATER HAVING FREE CHLORINE RESIDUAL

The system is all surface water and, thus, does not blend with water containing a free chlorine residual. They wholesale water to some small groundwater systems but have no knowledge of problems that they encounter with blending.

The staff gets a number of complaints from aquarists who do not understand the problems associated with dechlorinating and de-ammoniating.

MEETING SUMMARY

AwwaRF Project No. 2847 **B&V Project No. 131832.100**

February 13, 2003

Participants: Utility A - Water Quality Manager
Utility A – Customer Services Manager
Mr. Alan Hess – B&V Philadelphia

RE: Site visit for chloramination project

General

The meeting was held to discuss the conversion in April 2001 to chloramines and operational issues associated with chloramination. Chloramination was started following construction of a new 16 mgd plant, which totally replaced an existing plant on the same site. The new facility utilizes the Actiflo Ballasted Flocculation process with alum and polymer coagulation, and includes five deep-bed GAC/anthracite/sand filters and one deep-bed GAC/sand filter. Chlorine is added at several locations in the process, with some added at the intake for preoxidation of iron and manganese and to assist the coagulation process, and the major disinfection dosage is added at the combined filter effluent The ammonia system utilizes one anhydrous ammonia storage tank, which is outside of the building, W&T V10K ammonia feeders and an injector system utilizing softened water, with discharge into a channel on the suction side of the high service pumps and after the backwash pumps, at a point about 2 hours of chlorine disinfection.

Chloramination Start-up

Ammonia storage and feed systems were included in the design of the new treatment plant. By approving the design of the new plant facilities, the state regulatory agency also approved the equipment required and therefore the use of chloramines. The regulatory agency did not require any special report or study prior to converting to the use of chloramines. A portion of the utility's distribution system extends into another state. The second state had no input, review or approval in the treatment process changes since they do not regulate the treatment plants outside of there own state boundary. As long as the distribution system water quality meets the regulations in the second state, the utility is able disinfect the water with chloramines.

The utility conducted an intensive customer information program prior to converting to chloramines. The program included: meetings with government leaders, local hospitals and medical centers; sending a letter to each commercial and residential customer with an information pamphlet entitled "What you should know about chloramines"; sending letters and holding an informational meeting for all doctors offices, kidney dialysis facilities and pet store owners in the service area; providing posters for pet stores and grocery stores; putting press releases in the local newspaper; calling every restaurant regarding concerns for live fish tanks; and held an in-house chloramine training workshop which every employee was required to attend. They also used information on what others had done to develop their program, included reviewing the literature, visiting other plants using chloramination and talking with their operators.

Start-up of the chloramination process was actually begun about two weeks following its announced start-up date to allow a period for chronically complaining customers to not impact the actual customer responses to the conversion. During this two-week period they received about 150 calls, mainly from customers with questions already covered in the informational pamphlet, and from some "experts" that thought ammonia was bad for humans. The customer service manager was very pleased with the success of their program and could not offer any suggestions for what they forgot to do or would have done differently. Specific training and SOPs were developed for the operators and laboratory for the new testing required to control the ammonia feed system and chloramination process. When start-up was finally initiated, it consisted of just starting the ammonia feed system.

Controlling Cl_2:NH_3 ratio

The chloramine start-up called for using a chlorine:ammonia ratio of 4:1, and they calculate their dosage as ammonia. The Water Quality Manager said that they are now operating at a ratio closer to 3.5:1. An amperometric titrator is used to determine free chlorine, monochloramine and dichloramine, which results in a reading of 0.1 mg/L or less of free chlorine being measured, which they always rounded up to 0.1 mg/L for reporting purposes. The January 2003 chloramine analysis data indicated that the average pH was 7.0 and the average total chlorine residual was 2.5 mg/L, with 0.1 mg/L free chlorine, 1.75 mg/L monochloramine, 0.5 mg/L dichloramine, and 0.02 free ammonia. The average ammonia dosage was 55 pounds per day. The plant lab analyses for mono- and dichloramine and free ammonia daily using both a Wallace and Tiernan Amperometric Titrator and the HACH Company DR4000 Spectrophotometer, and adjust the ammonia dosage off the Clearwell effluent free chlorine residual. They have not had any problems controlling the ammonia dosage using this technique.

Annual Distribution System Conversion to Free Chlorine

The system chose to make an annual conversion to free chlorine for two weeks in the spring of 2002 in conjunction with their annual flushing program. They had not experienced any problems with the chloramine in the distribution system, but did it because "other utilities have done it", and it is recommended in the AwwaRF Chloramination Manual. The utility did not notify anyone that they were making this switch, since they didn't feel that it would affect anyone or cause any problems since consumers had already installed whatever they needed for removal of chloramines. They simply turned off the ammonia feeders, and used 2.3 –2.5 mg/L free chlorine residual instead of a chloramine residual, maintaining a minimum residual of about 0.7 mg/L in the far ends of the distribution system. At the same time, they flushed the distribution system from the water treatment plant to the ends of the system, with the flushing crews sampling for free chlorine residual.

They had no unusual customer complaints during this conversion, except for dirty water complaints related to flushing. They only had about 5 customer complaints regarding the chlorinous taste in the water. They do not have any documented evidence of any benefits from the conversion to free chlorine, however they have only done it one time, and are not sure if they will do it again.

Water Quality

While the chloramine eliminated the free chlorine odor in the finished water, it may have made other odor more noticeable, such as MIB and Geosmin The utility experienced a very bad taste and odor episode in November –December 2001. The problem was much worse than normal (raw water MIB concentration from 150 – 300 ppb in areas of the watershed), and all treatment facilities along the River had difficulties in treating it. In general the utility does not believe that the switch to chloramines has increased the number of T&O episodes, but rather that their customers do notice them a little earlier. This has changed their approach to monitoring and treatment of T&O problems. The new plant has a limited PAC treatment capability with the Actiflo process, and the filters only have 36 inches of GAC. They do not measure free ammonia in the distribution system, but do monitor nitrates and nitrites on a monthly basis at several locations. They do not have any problems with HPCs or coliforms in the distribution system, and maintain a minimum total chlorine residual of 0.4 mg/L in the summer months, where they were only able to maintain 0.02-0.04 mg/L when using free chlorine in the distribution system. They have reduced the Running Annual Average THMs from 65-81 ug/L down to 52 ug/L since they started using chloramines. Estimated average residence time in the distribution system is 7 days, with a maximum of 10 days (this is higher than the 1 day for each in the original survey response). They have not experienced any problems with nitrification, or any indication of conditions that would preclude this problem.

Utility A - Anhydrous Ammonia Storage Tank

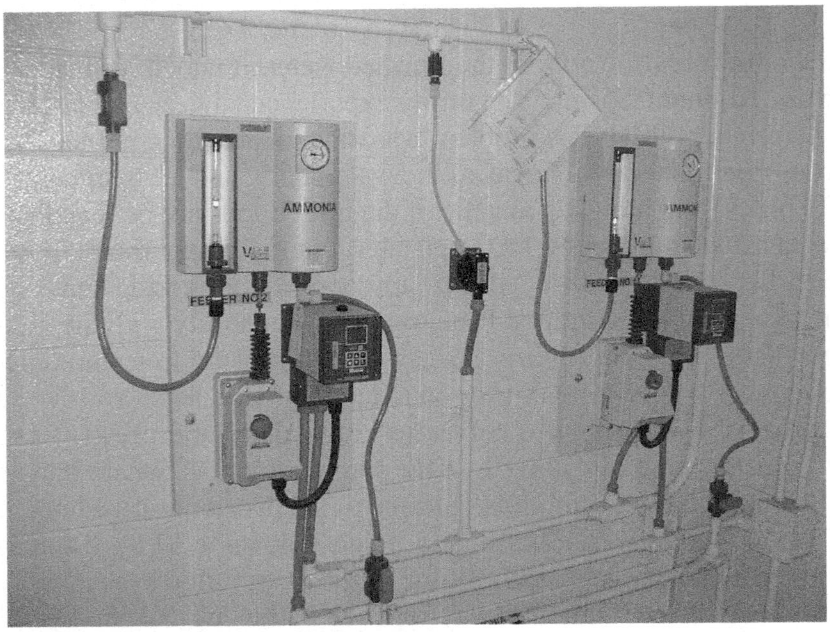

Utility A - Ammonia Feeders

Utility A – Ammonia Injector Water Softening System

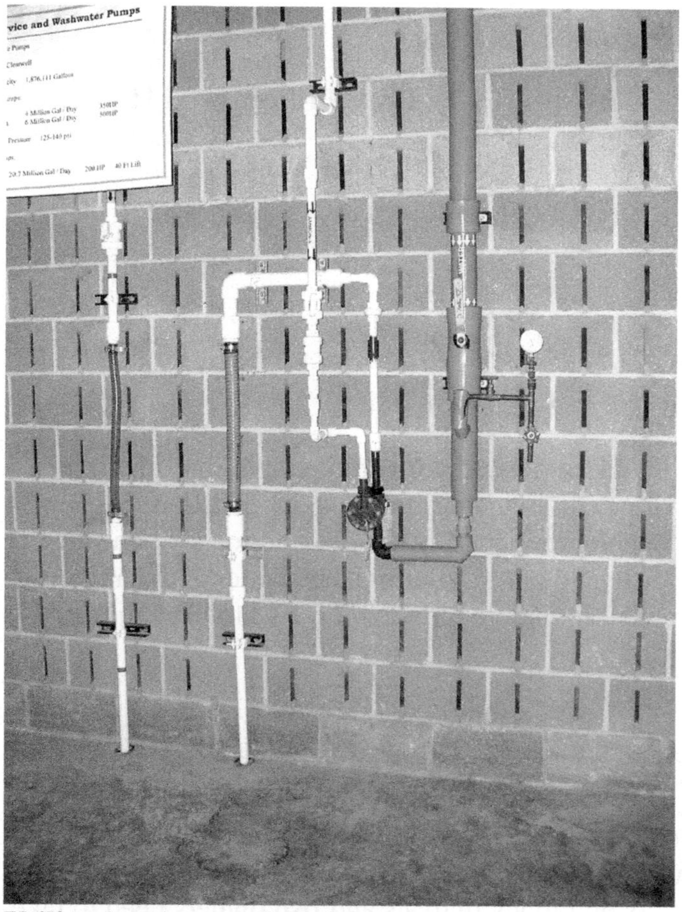

Utility A – Ammonia Injector

MEETING SUMMARY

AwwaRF Project No. 2847 **B&V Project No. 131832.100**

February 19, 2003

Participants: Utility B - Manager, Treatment & Quality Control
Alan Hess – Black & Veatch

RE: Site visit for chloramination project

General

The meeting was held to discuss operational issues associated with chloramination. The utility owns and operates 9 water treatment plants with a combined design capacity of 126 mgd, as well as a number of well sites. They have practiced chloramination in the distribution system since the 1970s, when it was added to improve the taste and odor of the finished water and maintain the residual in the system. It has served them well in maintaining low levels of DBPs in their distribution systems, and for the control of biofilms.

Conversion of Systems to Chloramines

The utility has acquired several water systems in the area, and has converted them from free chlorine to chloramines as soon as possible, based on their successful experience with using this type of residual in their major supply sources of the past 30 years. When these systems are converted, the utility sends a letter to each customer explaining what they are doing and why they are doing it, the benefits and that customers on kidney dialysis and fish owners need to take special precaution and pretreat the water before using it for these purposes. They also issue press releases in the local papers announcing the change and the potential impacts of it on the customers.

They personally contact, using employees in the local water system whenever possible, local hospitals, dialysis units, clinics and businesses about the conversion and explain what will happen and any potential impacts that they may experience. They also provide a poster for use in pet stores explaining the need to take the chlorine out of the water added to fish tanks.

The state regulatory agency does not object to the conversion of a system to chloramines as long as all the paperwork for the change is filed for conversion to chloramines, and the regulatory agency does not get involved in the chloramination implementation. However, the utility does have to apply for a permit for the new equipment and change in type of disinfectant. The regulatory agencies Regional Sanitarian or Engineer inspects the equipment when it is all installed and issues an Operating Permit. This also requires some changes to the plant O&M Manual, RMPs and Emergency Response Plan.

Nitrification

During 1994 the utility experienced some nitrification problems in some of their distribution system storage tanks during the winter. The determined that some highway departments were using urea as a deicing agent on the roads, which resulted in a higher level of ammonia in the source water, and also that the tanks were not being turned over quickly enough. Other sources of note in winter months were from the increases in levels in the raw water from the relaxing of the limits on wastewater plants for ammonia during the winter months.

Controlling Cl_2:NH_3 ratio

The utility uses a ratio of 4:1 as ammonia, and has no trouble maintaining this ratio. They monitor for free ammonia as part of their control practice. They try to maintain 0.1- 0.2 mg/L free ammonia in the treated water.

Operation

The utility had some problems with the use of anhydrous ammonia in the past, mainly on maintenance of the system and regulatory compliance issues, and now uses aqua ammonia at their surface water plants and ammonia sulfate at their well and distribution system interconnection feed installations. The aqua ammonia is fed with a pump with an automatic degassing head. The storage tanks are usually installed outside of the building, and some of them have been shaded or have a system installed so that they can be sprayed with water to reduce off-gassing. The ammonia sulfate is mixed to a 2-½ % solution, and fed with positive displacement pumps.

Blending

The only blending that they have is where water from their groundwater systems, which use free chlorine, mixes with water from their chloraminated surface water systems. They receive occasional customer complaints related to taste and odor and temperature changes. Occasionally, when the groundwater systems cannot meet their demand, surface water is added into these systems to supplement the supply.

Rechlorination

Rechlorination is practiced in about 20 locations in the utility's various distribution systems. This is controlled by using a chlorine residual analyzer with a compound loop control system. Chlorine is added to the water, which already contains ammonia from the decay of the chloramine residual, to return the water to the desired total chloramine level.

Water Quality

If they are experiencing problems with HPC or Coliforms in an area of the distribution system, they will isolate the area affected and use short periods of free chlorine addition into the system, injecting it through a hydrant and flushing it out of a hydrant at the far end of the area. They normally add 2 -3 mg/L free chlorine for about 8 hours in this process.

In 2002 the areas requiring this treatment were mainly those where flow reversal occurred as the source of supply switched during an extreme drought condition. They had samples with atypical organisms, and used this free chlorine burnout to clean the areas affected.

Problems with low chlorine residual are usually seasonal, and normally occur in the same locations. They recognize a need to improve distribution system flow patterns, especially in areas where wells are shut down for one reason or another.

They monitor for free ammonia, nitrate and nitrite in the distribution system, mainly at storage tanks.

The state regulatory agency has required the utility to provide a dechlorination system on some of their storage tank overflows. They have completed this on one tank so far, using an ascorbic acid tablet system, which will not hurt children or animals if they somehow come in contact with the chemical.

Disinfection
The utility follows the AWWA guidelines for disinfection new or repaired distribution system pipelines. Two samples are collected 24 hours apart for bacteriological testing and turbidity. For storage tanks, they also do a VOA test, odor and let the free chlorine residual decay in the tank to a low level, then fill the tank with chlorinated water from the distribution system.

Free Chlorine Switchover
The utility is evaluating the benefits to water quality of an annual switchover to free chlorine. They did this for a 2-week period in one of their systems in 2002 while they were doing some maintenance on the ammonia system equipment. They did not notify anyone of the switchover, treating it as a maintenance related practice. The switchover resulted in an increase in customer complaints with regard to the chlorinous taste and odor of free chlorine, even though the system was only converted to chloramines in the winter of 2001. The chlorine decay rate increased during the switchover, both through the plant clearwell and through the distribution system, with values of 0.1 mg/L measured at the far ends of the system. Since low chlorine residuals are a potential cause for a security concern, the utility decided not to have a switchover in any of their systems this year.

Recommendations for a Successful Chloramination Program

- Good planning
- Sell the advantages to Utility Management, Regulatory Agency and the Consumers (Local Government officials, businesses, and residents)
- Select good equipment – don't shortchange or compromise yourself
  - Designed for the product being used (don't reuse equipment just because you have it if is not truly compatible)
  - Use proper tank sizes (sized for full delivery) (if they heat up too much you can get degassing)
  - Pipelines (if suction line diameter is too large you can get degassing)
  - Valves
  - Feeder/Pumps (degassing design)
  - Injectors (provide ease of removal for cleaning scale buildup)
  - To keep degassing down order a lower percentage aqua ammonia
- Carefully select where the ammonia will be added in the process, and evaluate any and all possible reactions and impacts
- Provide training for everyone on: equipment, process and PR information

MEETING SUMMARY

AwwaRF Project No. 2847 **B&V Project No. 131832.100**

February 13, 2003

Participants: Utility C - Chemist
Utility C – Production Supervisor
Mr. Alan Hess – B&V Philadelphia

RE: Site visit for chloramination project

General

The meeting was held to discuss operational issues associated with chloramination. The utility operates three water treatment plants, with a total capacity of 84 mgd, and an average 2001 production of 45.2 mgd. One plant has used chloramines since it was built in 1974. In the 1990's changes were made in the distribution system flow patterns, which resulted in blending of the waters from this plant with water from another plant in a tank that serves a portion of the distribution system. Problems with maintaining a chlorine residual in the distribution system served by this tank resulted which were caused by mixing free chlorine water with chloramine water. Once this was realized, the utility implemented plans to convert the second plant to chloramines, and the system was started in the summer of 2000.

Initial Conversion to Chloramines

There were no special requirements for the utility when they decided to convert the second plant to chloramines, especially since they were already using this technology at another plant. They did a public notification, but concentrated on notification through the medical community and the Kidney Foundation. The number one priority was dialysis centers and hospitals, and they had developed a good relationship with them from using chloramines at the other plant. Home dialysis users were a concern, but the Kidney Foundation took care of notifying all of these users. They used the yellow pages in the phone book and called pet shops, restaurants, doctors offices, manufacturing plants, and other potentially impacted customers. One customer with a gold fish pond was impacted, and they ended up buying some replacement gold fish for him.

The response to survey question 6 on significant start-up problems should have included the fact that the increased iron levels may have been related to them also changing the corrosion inhibitor being used at the same time (Zinc Phosphate to Polyorthophosphate)

Controlling Cl_2:NH_3 ratio

The target Cl:NH ratio is 4:1, and control of the Cl:NH ratio is accomplished by decreasing the ammonia feed rate until they see a drop in the chlorine residual. Then they set the dose back to the previous (pre-decrease) dose. They do this after every major pumpage change in the plant, which only happens 1 or 2 times per month. They recommend controlling the ammonia feed so that no excess ammonia is added.

Operation
The plant uses 20 percent Aqua Ammonia, which is stored in a bulk storage tank. BIF Series 44 metering pumps are used to feed the ammonia from a day tank. The whole storage and feed system is in the same room in the lower level of the plant. Leaks in the storage, day tank and feed system were very noticeable in the room. A separate exhaust system exists for this room, and the door from the plant into the room is kept closed at all times to prevent the fumes from entering the rest of the building.

Water Quality
The system has not had any history of nitrification or any other problems related to use of chloramine. They do not measure free ammonia in the distribution system, but do monitor for nitrate and nitrite at a higher frequency during the warmer weather months. Chlorine residuals are now all higher throughout the distribution system. Customer complaints related to chlorinous odors are lower, and distribution system THMs are all lower (reduced from 40 ug/L to <30 ug/L).

Disinfection
Disinfection is provided by a 1.5 mg/L dose of chlorine to the raw water for oxidation of iron and manganese, and sufficient chlorine is added to the filter effluent to provide 2.0 mg/L free chlorine at the clearwell effluent, where the ammonia is added. Because the raw water is of very high quality and has a low DBP precursor level, they do not form significant THMs through the treatment plant.

Annual Conversion to Free Chlorine
The older plant has practiced an annual conversion to free chlorine since the 1970s. The conversion is now coordinated so that both plants do it at the same time each year. Originally the system switched to free chlorine for the winter season, before THMs were recognized as a problem. In the early 1990s they reduced this to a 2 week period each winter. They make the switch in the winter because the older plant does not have a THM problem if it is done during this time of the year. The utility made the decision to continue the period of annual free chlorine residual as a precautionary measure, providing a time of higher disinfection.

Conversion is made by just turning off the ammonia feed pumps. They do not notify their customers that they are making any changes, but do put a message on their answering machine and inform the state regulatory agency that they are making the change. They use this period to perform any required maintenance, repairs or modifications to the ammonia system. In 2003 they are replacing the feed piping to the application point.

They have no documented benefits of the annual free chlorine conversion, other than the absence of any long-term distribution system problems. The only customer complaints related to the annual conversion are from customers who are not accustomed to the free chlorine odor in the water. These were described as "very vocal" during this year's switchover, even though it was done in February when the water was very cold.

Utility C - Ammonia Storage & Day Tank

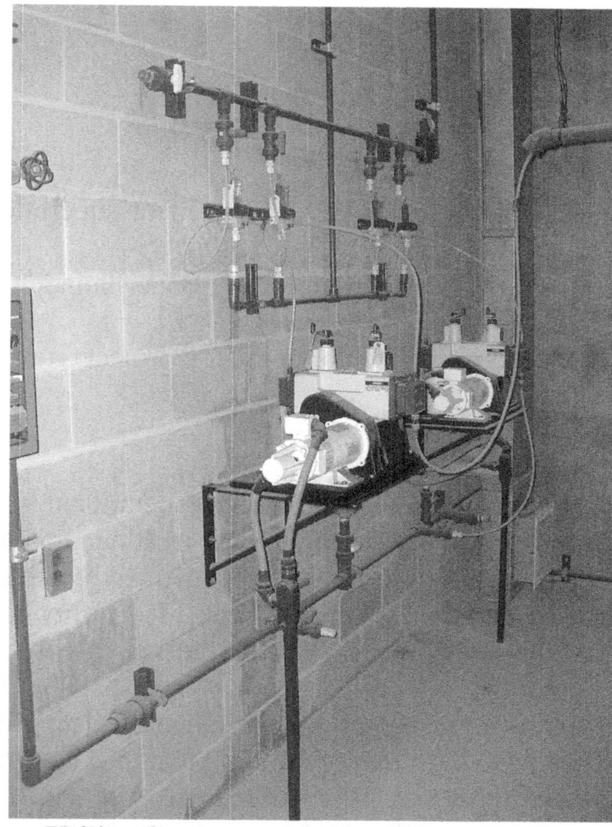

Utility C - Ammonia Feed Pumps

MEETING SUMMARY

AwwaRF Project No. 2847

B&V Project No. 131832.100
February 14, 2003

Participants: Utility D – Filter Plant Superintendent
Utility D - Chemist
Mr. Alan Hess – B&V Philadelphia

RE: Site visit for chloramination project

General
The meeting was held to discuss operational issues associated with chloramination. The utility operates a 30 mgd surface water treatment plant, withdrawing water from the combined flow of two branches of a Creek, supplemented by discharges from the two lakes, located upstream on one of the branches of the Creek. The utility has provided chloraminated water in the distribution system for over 65 years.

They do not have any literature that they supply to new customers or businesses on chloramines. They have been using it so long that they assume that everyone knows what to do. The only problem that they have had recently was with a new aquarium owner who was overdosing the dechlorinating chemical and depleted the dissolved oxygen level in the water, which resulted in the fish dying.

Nitrification
They have experienced isolated problems in the last four years in small areas of the distribution system where there is a long main feeding a few (15-20) houses on an old, unlined cast iron main. They have not had any problems during cold weather, but during warm weather they have had some red water problems. They have maintained a detectable chlorine residual (0.1-0.3 mg/L) at the end of the distribution system. They used a Hach BART test kit to look for nitrifying bacteria, and it showed that a low level of them were present (light pink range on test). Nitrate and nitrite tests showed a low level (0.008 mg/L). They suspect that the problem is related to increasing the distribution system pH higher (to 8.0) for compliance with the Lead and Copper Rule.

Controlling Cl_2:NH_3 ratio
They use a 3:1 to 3.5:1 ratio as ammonia. Ammonia dosage is based on the filter effluent free chlorine residual, with the goal of a trace (0.01-0.02) mg/L free chlorine entering the Filtered Water Storage basin following the ammonia addition. They do not currently measure free ammonia residuals in the plant or distribution system, but plan to start monitoring this parameter in 2003. They maintain a 1.2 mg/L total chlorine residual leaving the Filter Water Storage Basins. They use an amperametric titrator for all chlorine residuals throughout the plant and distribution system. All samples are analyzed in the plant laboratory, with distribution system samples put on ice when they are collected.

Operation
The ammonia system utilizes one 500-gallon anhydrous ammonia storage tank, which is located outside of the building. The storage tank is on a load cell to monitor the storage quantity. The storage tank equipment is provided by the ammonia vendor, and must be changed whenever a new vendor is contracted. This requires careful coordination to empty, remove and install the new tank, which at times requires the use of a temporary storage tank. Since chemical supply contracts are on a calendar year basis, this normally occurs in January. The ammonia system is direct feed, with the 40-psi ammonia gas fed to 2-100 ppd W&T Ammoniators, and then into the filter effluent conduit. The ammonia injection tubes are removed and the ends dipped in acid to remove calcium carbonate deposit buildup about every two months. They notice a decrease in feed capacity as an indication that the tubes are starting to scale up.

During periods of sustained (several days) of below freezing weather, they can experience a decrease in the pressure of the ammonia storage tank, which has dropped below 20-psi on occasion. This reduces the maximum capacity that they can feed. The tank is located where it has a southern exposure, and to reduce this problem, they cover the storage tank with a black rubber sheet to absorb sunlight, and under extreme conditions put a tarp over the tank and circulate warm air around the tank to recover the feed capacity. They do experience icing on the tank from December through March, particularly during winter rain events.

They believe that they have one of the few direct feed ammoniation systems in the country, and are concerned that the supplier (W&T is the only one still making them) will discontinue manufacture of this design ammonia feeder. They have recently installed two new direct feed Ammoniators to make sure that they have good equipment for the foreseeable future.

Blending
They sell water to some small systems that use free chlorine. They do not know how these systems deal with the difference in residual type or if they have had any problems. They have not had any complaints of problems from these systems.

Water Quality
The system has very low DBP levels throughout the distribution system (THMs 25-30 ug/L, HAAs <15 ug/L), as well as stable chlorine residuals. They rarely have any taste and odor complaints, and get more calls about air in the water that gives it a cloudy appearance when the glass is first filled.

Booster Chlorination
They rechlorinate the water at one pump station location in the distribution system. This is a small system, and they add 0.5 mg/L of residual to the water leaving the station, to increase the residual from about 0.4 mg/L to 0.9 mg/L. They use a positive displacement pump and injector feed system, and a 15-gallon sodium hypochlorite storage tank (about 1 week supply), all contained in a Rubbermaid Shed inside the pump station. The feed pumps are on/off with the pump station pumps.

Utility D - Anhydrous Ammonia Storage Tank

Utility D - Ammoniators

MEETING SUMMARY

AwwaRF Project No. 2847 **B&V Project No. 131832.100**

February 12, 2003

Rev. April 18, 2003

Participants: Utility E, Chief operator
Dr. Lee Harms, B&V – Kansas City

RE: Site visit for chloramination project **Location:** Utility E Water Treatment Plant

General

Utility E has a single 26 mgd surface water treatment plant plus three wells that feed water without treatment, except chlorination, into the distribution system. The water quality of the two sources is considerably different, as indicated below:

- The surface water is very low hardness and alkalinity (30-35/35) while the ground water is low hardness but high alkalinity (11-15/200). All three wells have slightly different water quality with alkalinities ranging from 165 to 280 mg/L.
- The surface water is disinfected with free chlorine followed with chloramines while the ground water supplies receive free chlorine only.
- The ground waters contain fairly high sulfur concentrations which make them objectionable to drink and generate most of the water quality complaints received by the Water Department.

Nitrification

Nitrification does occur periodically as evidenced by low chlorine residuals. A chemical plant within the service area is closely monitored for chlorine residual and when the chlorine residual starts to drop at this location it often is an indicator that nitrification is beginning. Chlorine residuals are routinely monitored at various locations throughout the system and when the drop in chlorine residual occurs, a free chlorine burn is initiated for 3 to 7 days. This procedure is repeated approximately every 6 months. They do experience some difficulty controlling the free ammonia which, although targeted at about 0.1 mg/L, sometimes rises to 0.3 to 0.6 mg/L. The plan is to purchase a Hach kit to more closely monitor free ammonia and monochloramine levels. Operators do test for free and total chlorine on all samples as another measure of controlling the ammonia. If greater than 0.1 mg/L free chlorine is detected in the treated water, the ammonia feed is adjusted.

Controlling Cl_2:NH_3 ratio

The ammonia feed rate is based on chlorine residual, not chlorine dose. Thus, the chlorine demand is taken into account. Target ratio is 3:1 to keep all the chloramine in the monochloramine form. The operator stressed the need to monitor ammonia and nitrate, especially when just beginning to use chloramines, as a means of educating operators so ammonia is not overfed.

Operation

Chlorine is added after the filters with ammonia added in the discharge from the clearwell. Aqua ammonia is used for the ammonia source, and some difficulty has been experienced when feeding the concentrated 29.4 percent solution. This solution vaporizes at warm temperatures (about 85° F) and can cause "air" gaps in the feed lines and the pumps can vapor lock. The operators have had to dilute this solution by hand in order to feed it correctly. Specifications are now written to receive a more dilute solution of aqua ammonia (19 percent) and this solution should not vaporize until about 120° F. Peristaltic pumps are used to feed this solution with the ammonia injected directly from a small pipe into the PVC pipe that exits the clearwell. No diffuser is used but good mixing occurs because of the turbulence in the pipeline.

When switching from chloramines to free chlorine they sometimes experience a sudden drop in chlorine residual. This reason for this drop has so far not been clearly identified. Dialysis at local hospitals is an issue and medical facilities should be notified when using chloramines.

Blending

Ground water disinfected with free chlorine is blended with surface water treated with chloramines. The wells are located opposite the treatment plant so some of the system receives only ground water but there is an interface which moves some in the system depending upon percentage of ground water being used and demand within the system. Very few water quality complaints are received from this interface region. Higher THMs are noticed in the region where the two source waters are blended. The chlorine residual in the ground water is normally about 2 mg/L of free chlorine and this is blended with the surface supply which has about 3 mg/L of chloramines.

Water Quality

The differences between the two sources have already been mentioned. Chlorine dioxide was tried at the wells to oxidize the sulfur and this worked well. Currently chlorine dioxide is not used. Water quality complaints originate primarily in the north and west portions of the distribution system and are primarily from hydrogen sulfide in the well waters. These complaints typically are found on dead ends of the distribution system and the situation is corrected by flushing the water line.

Some hydrogen sulfide is noticed at the wells but most of the hydrogen sulfide dissipates before it reaches any customers. Some of the hydrogen sulfide is removed at the well facilities through aeration fans and chlorination in the ground water storage tanks.

THM levels typically increase only about 10 μg/L after the water leaves the treatment plant. The WTP is operated to produce water with a pH of ≤ 7.3 to avoid exceeding the THM limit. After the water leaves the WTP ground storage tank additional chlorine, ammonia, and caustic are added. The desired pH of the water leaving the plant is 8.3. HAA5 have been running about 40 μg/L.

Disinfection

Disinfection is achieved using free chlorine after the filters and through the clearwell. A new baffled clearwell is being constructed. Typical chlorine residuals leaving the treatment plant are

<0.1 mg/L free chlorine and 1.7 to 2.6 mg/L total chlorine (chloramines). Residuals at the wells run about 1.8 to 2.0 mg/L as free chlorine, and residuals at the interface of the two supplies typically are about 0.6 to 0.7 mg/L free chlorine and 1.2 to 1.6 mg/L total chlorine.

MEETING SUMMARY

AwwaRF Project No. 2847 **B&V Project No. 131832.100**
March 27, 2003

Participants: Representative, Utility F
Mr. David Hoot, B&V

RE: Site visit for chloramination project **Location:** Utility F WTP

General
Utility F has a single 16 mgd water treatment plant, with raw water provided by groundwater supply from 12 wells (7 wells at the WTP and 5 offsite wells) drilled into a large aquifer. Utility F has an average demand of approximately 24 mgd, and obtains additional water supply for its customers by purchasing finished water with 8-10 connection points at various locations within the City. The water distribution system has approximately 500 miles of pipe.

The water treatment process consists of lime softening, filtration and disinfection, followed by the addition of fluoride, with no major operational concerns identified. Chorine disinfection is accomplished with the use of chloramines. Monthly, 134 different locations throughout the service area are sampled for bacteria, chlorine, cloudiness and iron.

Nitrification
Nitrification is not a significant concern within the water system, especially the WTP. There is limited evidence that nitrification occurs in the outer reaches of the distribution system as depicted by periodic low chlorine residuals (approx. 3-5 tests may fall below 0.6 mg/l). Utility F maintains a flushing program, and performs a free chlorine burn once per year, in the fall for two weeks. The chorine residual average appears to increase during the winter months (Dec.-Feb.), and then falls off during the summer to a lower level approaching September, prior to the burn. HPC's and biofilms are not a problem that has been observed. Typically DBP's average 20, (DBP are reported to Health Department Quarterly) (The Health Department does not have a specific requirement for testing DBPs during free chlorine flushing). Utility F does not test or measure for ammonia in the distribution system, but take tests/measurements every 4 hours at the WTP.

Controlling Cl_2:NH_3 Ratio
The target Cl_2:NH_3 ratio is 5:1 based on a chlorine residual goal of 3.5 to 4.0 and they normally achieve this through balanced ammonia and chlorine feed rates within the WTP. Ammonia gas is fed into the system prior to the lime softening treatment process, and then the chlorine gas in added. The time from ammonia addition to chlorine addition is approximately 40-50 minutes. Utility F has had limited problems balancing or maintaining their ratio (maybe 1-2 per year), sometimes associated with/after their yearly free chlorine burn cycle.

Operation
No significant/major operational problems or issues were noted. Utility F is currently planning on replacing the existing ammonia feed system with a new on in the near future. The WTP is also undergoing a plant expansion to include membrane treatment. The WTP is currently operating at capacity on a continuous basis.

Blending
They currently do not blend ground water and surface water.

Water Quality
The only water quality issue mentioned by Utility F involved a slight color concern coming from their groundwater. It is a borderline concern and does not impact the finished water to any major extent. Utility F has had no problems meeting the Total Coliform Rule, and has never tested positive for coliforms. Free chlorine is used for new and repair pipe service testing following AWWA standards. Utility F typically requires the contractor to take two test samples for the Health Dept. and they take two verification samples for internal tests. Utility F experiences some periodic water quality degradation at the outer reaches of the system and on dead-end pipes and utilizes a flushing program or automatic flushers to keep the system fresh.

Disinfection
The chlorine and ammonia are added to the system at different locations within the plant, with the chlorine being added before the clear well and filters. There is sufficient contact time in the clear well to obtain their disinfection, and appear to have no problems meeting disinfection levels.

MEETING SUMMARY – Utility G

AwwaRF Project No. 2847 **B&V Project No. 131832.100**

Participants:

Senior Program Manager, Utility G
Engineer, Utility G
Dr. Mark Waer – B&V, Phoenix

RE: Chloramination Project Site Visit

Date: February 7, 2003, from 8:45 to 9:45 a.m.

Location: Utility G Water Treatment Plant

INTRODUCTION

This meeting was held to discuss operational issues associated with chloramination of drinking water for disinfection. I met with Utility G's treatment engineer, to discuss these issues.

Utility G operates three water treatment plants. Utility G's engineer had excellent knowledge of the operation of all three facilities.

DISINFECTION

WTP 1. WTP 1 receives its water from two lakes through transmission lines of 3 and 18 miles length. Free chlorine is added at the intakes and disinfection is achieved through the raw water transmission lines. When the water reaches the plant it contains between 0.5 and 1.5 mg/L of free chlorine. Ammonia and additional chlorine are added at the flash mix. The water is coagulated, flocculated, settled and filtered. Additional chlorine and ammonia may be added at the end of the plant. Pre-ozone is currently being installed at this facility.

WTP 2. Primary disinfection is achieved with pre-ozone. Chlorine and ammonia are added at the flash mix, downstream of the ozone contactors. The water is then coagulated, flocculated, settled and filtered. Additional chlorine and ammonia may be added at the end of the plant.

WTP 3. In spring, summer, and fall, chlorine and ammonia are added at the flash mixer. The water is coagulated, flocculated, settled, and filtered. Additional chlorine and ammonia may be added at the end of the plant. Primary disinfection is achieved with chloramines through the entire treatment plants.

During winter, the contact time through the plant is insufficient to maintain primary disinfection with chloramines, so the point of ammonia application is moved to downstream of the flocculators. Additional chlorine and ammonia may be added at the end of the plant. Primary disinfection is accomplished with free chlorine under these circumstances. Ozone facilities will be completed in about 2 years.

CHLORINE TO AMMONIA RATIO

Essentially the ammonia feed is held constant at 9 to 10 pounds per million gallons (1.1 to 1.2 mg/L) at the flash mix and about 25 pounds per million gallons (3 mg/L) chlorine is added at the intakes and at the flash mix. This amounts to a ratio of about 5:1 based on dosage. They adjust chlorine for residual but ammonia remains nearly constant.

They have no current feed problems or balancing problems. They do not form dichloramines. A few years ago, they installed new ammonia facilities which gave them more storage, more feed capacity, and better feed control. Since that installation, they have had no problems with their system. The additional storage capacity resulted in a lower chemical price also.

NITRIFICATION

About 11 years ago, the system had nitrification problems. This was detected by loss of chlorine residuals through the distribution system. They found that by increasing the chloramine residual leaving the plants to a minimum of 3.0 mg/L and by cleaning water tanks in the distribution system, they have been able to control nitrification in their system. Their tank cleaning procedure is to empty the tanks, remove any sediments that may have accumulated, and disinfect the tank with chlorinated water. The disinfection water is then wasted.

They do not have a program of burning the system with free chlorine. They currently monitor the distribution system for nitrite and the plant for free ammonia. They also sometimes check the distribution system samples for free ammonia.

OPERATIONS ISSUES

They currently have no operations issues connected with chloramine usage. As the next two plants convert to pre-ozone, they will change their chloramine operations to reflect this change.

WATER QUALITY

Sometimes they need to add more chlorine and ammonia at the post to achieve the minimum 3.0 mg/L leaving the plant.

BLENDING WITH WATER HAVING FREE CHLORINE RESIDUAL

Utility G provides water to other communities, but they all are 100% surface water and do not blend with another source water containing free chlorine.

MEETING SUMMARY – Utility H

AwwaRF Project No. 2847 **B&V Project No. 131832.100**

Participants:

Water Quality Manager, Utility H
Assistant Water Systems Superintendent, Utility H
Dr. Mark Waer, B&V-Phoenix

RE: Chloramination Project Site Visit

Date: February 7, 2003, from 8:45 to 9:45 a.m.

Location: Utility H Water Treatment Plant

INTRODUCTION

This meeting was held to discuss operational issues associated with chloramination of drinking water for disinfection. I met with the Water Quality Manager and the Assistant Water Systems Superintendent to discuss these issues.

The Water Department operates four water treatment plants. The utilities representatives had knowledge of the operation of all of the plants.

DISINFECTION

WTP 1. In spring, summer, and fall, chlorine and ammonia are added at the flash mixer. The water is coagulated, flocculated, settled, and filtered. Additional chlorine and ammonia may be added at the end of the plant. Primary disinfection is achieved with chloramines through the entire treatment plants.

During winter, the contact time through the plant is insufficient to maintain primary disinfection for viruses with chloramines, so the point of chlorine application is moved to the plant influent. This provides a short period of free chlorine residual for disinfection of viruses. Ammonia is then added at the rapid mix. Additional chlorine and ammonia may be added at the end of the plant. Primary disinfection is accomplished with free chlorine and chloramines under these circumstances. Ozone facilities have just been completed and are in startup at the time of this meeting.

WTPs 2 & 3. At plants 2 and 3, chlorine and ammonia are added at the head of the plants and primary disinfection is achieved with chloramines year round through the contact time within the plant. The water is coagulated, flocculated, settled and filtered. Additional chlorine and ammonia may be added at the end of the plant.

WTP 4. Primary disinfection is achieved with pre-ozone. The water is then coagulated, flocculated, settled and filtered. Chlorine and ammonia are added after the filters, to allow biological filtration.

CHLORINE TO AMMONIA RATIO

The chlorine to ammonia ratio is held between 4 to 5:1 based on chlorine dosage. They feed ammonia to maintain zero free chlorine residual and no formation of dichloramine in the distribution system. Currently, there have been some upsets to the ratio when the new plant modifications have been started up, but these are just the product of getting used to new modes of operation.

NITRIFICATION

As recently as 6 years ago, the system had nitrification problems in the neighborhood of finished water storage tanks in the distribution system. This was detected by loss of chlorine residuals in those areas of the distribution system. They found that by minimizing detention time in their storage tanks through aggressively exercising the tanks, they have been able to control nitrification in their system. They have not had any problems with nitrification since they instituted that program of exercising the tanks.

They do not have a program of burning the system with free chlorine. They currently monitor the distribution system for nitrite and occasionally for free ammonia when a feed problem is suspected. HPCs are usually less than 100 cfu/mL.

OPERATIONS ISSUES

They currently have no operations issues connected with chloramine usage. As WTP 1 converts to pre-ozone, they will change their chloramine operations to reflect this change (i.e. move application point to post filtration to allow for biological filtration.

WATER QUALITY

The only plant upset that they mentioned was some atrazine concentrations from a reservoir that had no impact on disinfection practice.

One problem that they mentioned was that water from the four plants has different water quality characteristics and mixes freely within their distribution system. The water quality is now becoming more consistent at the plants resulting in fewer problems with mixing.

BLENDING WITH WATER HAVING FREE CHLORINE RESIDUAL

The system provides water to several other communities, but they all are 100% surface water and do not blend with other source water containing free chlorine.

MEETING SUMMARY

AwwaRF Project No. 2847 **B&V Project No. 131832.100**

February 11, 2003

(revised March 19, 2003)

Participants: Plant Superintendent, Utility I
Dr. Lee Harms, B&V – Kansas City
Mr. Gary Smith, B&V

RE: Site visit for chloramination project **Location:** Utility I WTP

General

The meeting was held to discuss operational issues associated with chloramination. Utility I operates four surface water treatment plants with a total capacity of 435 mgd plus 99 ground water plants that disinfect water from 202 wells. Most of the ground water facilities use free chlorine but a few now use chloramines. Not all ground water facilities mingle water with surface water. They operate 12 interconnected water systems and of these 12, only the main system receives surface water. The Main system also receives water from about 70 of the ground water plants. Other plants make up much smaller water systems, only one of which serves greater than 10,000 people.

Nitrification

Utility I is aware that they sometimes experience nitrification within the distribution system. This is primarily ascertained by monitoring chloramine residuals. If the chloramine residual drops in a certain area, a crew is sent out to check the distribution system. Most of the time (90 percent as an estimate) a valve has been closed and the area with the low chloramine residual has become isolated and stagnant. The situation is then corrected by opening up the valve and letting fresh water into the area.

If a physical reason for the low chlorine residual cannot be determined, the suspected main will be traced from a region of satisfactory residuals to where the residual becomes too low. Samples are collected and brought back to the lab where tests are conducted for nitrite, nitrate, ammonia, and nitrifying bacteria. These tests are not done routinely but are done if a problem is suspected. If the area of low residual is localized, a portable chlorine trailer is brought in to add free chlorine to just the affected area. If the area of low residuals is too large to be corrected using the trailer, the treatment plants switch to free chlorine, which is usually a last resort to correcting the problem. If practiced, the free chlorine burn runs for 7 to 10 days at a free chlorine dose of about 3 mg/L. Benefits from the free chlorine burn are that the chlorine residual comes back up, and the nitrate and nitrite levels drop back down to approximately what they are leaving the treatment plants. Free ammonia at the plant discharge is <0.2 mg/L; routine monitoring of ammonia within the system is not practiced.

Going to a free chlorine burn using all the surface water plants can be an annual occurrence but sometimes they can go several years between burns without seeing nitrification within the distribution system. One industrial customer has a low tolerance for THMs in their bottled product and going to free chlorine throughout the system causes them difficulty, and they have to use more carbon to prepare the water in their bottling process.

Controlling Cl_2:NH_3 ratio

The Cl_2:NH_3 ratio target is 4:1 and is based on dose; chlorine demand is not considered because the chlorine and ammonia are added within seconds of each other. Plans are being finalized to install monochloramine/ammonia monitors (Hach Chemical Co.) that will allow the operators to more accurately control the process. The operators set the chlorine dose to maintain the residual they want, then the ammonia is adjusted to ensure conversion to monochloramine. Target free ammonia after mixing is 0.1 to 0.2 mg/L.

Operation

Major difficulties with the chemical feed systems have not been experienced. They do not use a diffuser to discharge the ammonia; it is added using a small diameter pipe pointed upstream ahead of the mixers and the large turbulent flow ensures adequate mixing. They recommend using a dedicated pump for each application point as they did experience problems attempting to control a split feed. They found that one basin might have chloramines while an adjacent basin may have free chlorine present. Operators should be schooled in doing both free and total chlorine residual tests so they can pick up on this situation.

Blending

Potential blending problems occur when chlorinated ground water is mixed within the distribution system with chloraminated surface water. A result of this blending can be taste and odor problems and/or a loss of chlorine residual. Utility I has only a few areas where ground water is introduced into pockets of surface water. In these areas they now add ammonia to convert the ground water disinfectant to chloramines also so they do not experience complaints from blending the two disinfectants.

Operators should be trained to understand chloramine theory and the breakpoint chlorine curve. The tendency when low chlorine residuals are reported is to increase the chlorine which can drive the chlorine residual even lower. Sometimes chlorine demand on the finished water is done so they can advise the operator of the maximum chlorine that can be fed without going through breakpoint chlorination.

Water Quality

Water quality is not an issue as far as chloramines are concerned. They do use chloramines to keep DBPs low. They add a polyphosphate sequestering agent at 12 ground water locations to sequester iron. They have a well equipped chemistry lab with which to monitor water quality. Enhanced coagulation has not been a problem and they have consistently removed > 35 percent TOC, which is their regulatory target.

Disinfection

Disinfection is accomplished by adding free chlorine just seconds ahead of the ammonia for virus inactivation (approved by State Primacy Agent) and then continuing to maintain a chloramine residual through the plant. They have a continuous CT calculation via their SCADA system which allows the operator to monitor CT. Some new and repaired mains do fail required coliform testing after construction but this is attributed more to poor construction techniques than the use of chloramines. They do use booster pumping stations to boost chlorine residuals at sites a long distance from the treatment plant (25 miles+).

MEETING SUMMARY

AwwaRF Project No. 2847 **B&V Project No. 131832.100**
March 27, 2003

Participants: Representative, Utility J Water and Sewer Dept.
Mr. David Hoot, B&V – West Palm Beach

RE: Site visit for chloramination project **Location:** Utility J WTP

General

Utility J operates and maintains a very large water treatment and distribution system for approximately 2.5 million people. They have three major water treatment plants in their system with a capacity of more than 467 mgd using chloramines, and a couple of small WTPs acquired from private utilities that use free chlorine. Utility J's raw water is obtained by groundwater supply from major wellfields associated with each plant, with a total of 78 shallow wells drilled into a large aquifer. Utility J provides approximately 25% of its finished water to various municipal customers in the metropolitan area.

The water treatment process consists of lime softening, recarbonation, filtration and disinfection, followed by the addition of fluoride, with no major operational concerns identified. Chorine disinfection is accomplished with the use of chloramines.

The site visit and observations were held at WTP 1, and the majority of the information provided relates to this plant, but it also applies to the other two major WTPs.

Nitrification

Nitrification is not a significant concern within the water system or any of the WTPs, and no nitrification problems have been observed. There is limited evidence that nitrification occurs in the outer reaches of the distribution system as depicted by periodic low chlorine residuals, and a slight increase in taste and odor complaints in August and September just prior to the system purge/burn. Utility J's distribution system is primary a looped system and they maintain a flushing and hydrant exercising program, and performs a free chlorine burn once per year, in the fall for two weeks. The chorine residual average appears to increase during the winter months (Dec.-Feb.), and then falls off during the summer to a lower level approaching September, prior to the burn. HPC's and biofilms are not a problem that has been observed. Typically DBP's are observed at WTP 2 only during the burn, with breakpoint chlorination related to the DBP problem. Utility J does not test or measure for ammonia in the distribution system, but take tests/measurements every 2 hours at the WTPs, and they sample for nitrates/nitrites quarterly.

Controlling Cl_2:NH_3 Ratio

The target Cl_2:NH_3 ratio is 6:1 based on a chlorine residual goal of 3.0 to 3.5 and a combined chlorine goal of 3.5 to 4.0. They normally achieve this through balanced ammonia and chlorine feed rates within the WTPs. Each WTP has varying ratios due to naturally occurring ammonia in the groundwater supply. The chlorine to ammonia ratios are as follows: WTP 1, 5.0:0.8 ppm (with the water having 0.5 ppm naturally occurring ammonia); WTP 2, 12.3:1.0 ppm; and WTP

3, 5.0:0.5 ppm (with both WTPs 2 and 3 having more naturally occurring ammonia). Combined ammonia is measured at all WTPs. At WTP 1, ammonia gas and chlorine gas are fed into the system right next to each other after to the lime softening treatment process. The time from ammonia addition to chlorine addition is approximately 10-15 seconds.

Utility J has had very few problems balancing or maintaining their ratio. They are in the process of changing from anhydrous ammonia to aqueous ammonia at all three WTPs this year for safety reasons.

Operation
Utility J has had no significant/major operational problems or issues related to chloramines or the feed systems, and plant and system operations are in good condition. Flows at all the plants are very consistent, and at WTP 1, flow requirements are increased/decreased in 10 mgd increments. WTP 3 is currently operating at capacity on a continuous basis.

Blending
They currently do not blend ground water and surface water. There is mixing of the groundwater supply from different wellfields depending on WTP and wellfield operational needs.

Water Quality
Utility J's water quality, both raw and treated, is generally very good and consistent. Color and VOCs are two areas of concern at the WTPs 2 and 3, respectively. One well field provides Plant 2 with water that has a color of 60, and another well field provides water with an elevated VOC content that requires air-stripping. Utility J has not had any issues with coliforms or meeting the Total Coliform Rule. Free chlorine is used for new and repair pipe service testing following AWWA standards, and the two TC test samples are taken and analyzed by Utility J personnel.

Disinfection
Chlorine and ammonia are added to the system at the same location within a 10 feet of each other, at the entrance to the recarbonation/sedimentation basin, before the filters and the clear well. There is sufficient contact time in the mixing basin, filters and clear well to obtain disinfection. Contact times for each plant vary, with WTP 1 having a 6-8 hour contact time, and WTPs 2 and 3 having 2-3 hour contact times. Some system modifications are being considered to increase contact times at plants 2 and 3. Utility J currently has no problems meeting disinfection levels.

APPENDIX F
EXAMPLE MATERIALS FOR PUBLIC COMMUNICATION

This appendix contains examples of several materials that could be adapted to communicate to the public the conversion to chloramines as a disinfectant. These materials were used successfully by Tampa Bay Water during their recent conversion to chloramines from free chlorine. Included within this appendix are samples of:

- Background Information
- A Comparison of Chloramines and Free Chlorine
- Special Information for Kidney Dialysis Patients and Centers
- A Letter Alerting Customers Regarding Changes in Water Treatment
- The Project Overview
- Questions and Answers
- A Glossary of Terms
- Special Information for Pet Stores and Fish Aquarium Owners
- A Pamphlet Entitled: Attention Fish Owners
- An Overview of a Public Information Program

2535 Landmark Drive, Suite 211, Clearwater, FL 33761-3930
Phone: 727.796.2355 / Fax: 727.791.2388 / SunCom: 513.7010
www.tampabaywater.org

TAMPA BAY WATER
Supplying Water To The Region

Tampa Bay Water Quality Project

Background Information

In 2002, Tampa Bay Water will stop using chlorine as its disinfectant residual and will begin using chloramines, a mix of chlorine and ammonia, in its treatment of drinking water. After initial treatment with a primary disinfectant, small amounts of a secondary disinfectant, or disinfectant residual, must be applied and maintained at all times within the distribution system to kill organisms that could make people sick. Tampa Bay Water will use chloramines as its secondary disinfectant and continue to use chlorine as its primary disinfectant.

While Tampa Bay Water already complies with all current water quality regulations, this change is being made in advance of new, stricter health standards expected to be imposed in the future under the national Safe Drinking Water Act. Chloramines were selected as the method of choice because they are the most cost-effective way of maintaining compliance under the new regulations. The use of chloramines is 20 percent less expensive than alternative methods of treatment, and at the same time reduces the formation of suspected cancer-causing compounds. And, by switching to chloramines, a more consistent water quality will be maintained throughout Tampa Bay Water's wholesale system.

The use of chloramines as a secondary disinfectant is not a new process and has been used for decades in many cities and counties across the country. In fact, 20-25 percent of the local governments within the United States currently use chloramines, and the number is expected to grow due to new regulations. Cities like Denver and Minneapolis have been using them since World War II. In Florida, chloramines are used in 30 utilities, including the Miami-Dade area of South Florida.

The change to chloramines will reduce the level of some regulated disinfection by-products formed when chlorine mixes with trace quantities of naturally occurring organic substances found in water. This is particularly significant because the use of chloramines will reduce the formation of trihalomethanes (THMs) and haloacetic acids (HAAs), two types of by-products suspected to cause cancer with prolonged exposure.

This change will not impact all Tampa Bay area residents in the same way. Any noticeable changes will depend on the type of final disinfectant each member government chooses to use and whether the chosen treatment differs from that member government's existing practices. A final disinfectant is necessary because the residual decreases with time and may disappear completely. For example, thecityof Tampa already adds chloramines to its water as a final disinfectant, so its customers will not notice a change. Hillsborough County will begin providing chloraminated water to residents of north Hillsborough County and will continue providing water treated with free chlorine to residents of south-central Hillsborough County, so only some of its residents may notice a change. Pinellas County, Pasco County and the city of St. Petersburg will begin treating their water with chloramines, so their residents may notice a change. New Port Richey, along with certain other areas that receive water from isolated sources, will not receive any chloraminated water from Tampa Bay Water, so those residents will not be affected. Municipalities that purchase water from Tampa Bay Water's members will be affected accordingly.

Those member governments that do switch to chloramines as a final disinfectant will need to make some equipment changes at their treatment facilities to allow for the addition of ammonia to their current treatment system.

Residents who will be receiving chloraminated water for the first time may notice a change in taste or odor. Chlorine will become much more difficult to detect. Other temporary impacts may be noticed in isolated areas as the chloraminated water is introduced into the member governments' retail distribution systems. Some customers may notice a temporary variation in water color as well as possible sediment in the water. These temporary impacts to water quality are similar to those experienced when a water main is replaced or routine maintenance is performed on the water distribution system and will cease as the system stabilizes.

Chloramines, like chlorine, must be removed from water before it is used for two specific purposes. Those purposes include:

- Kidney dialysis treatments and
- Keeping live fish, such as in an aquarium or pond

Since the process for removing chloramines is different from some of the methods used to remove chlorine, medical facilities and dialysis units, as well as pet shops, seafood merchants, restaurant managers and area attractions featuring marine life will be notified in ample time to prepare for the conversion. (For more information please see inserts entitled *Special Information for Kidney Dialysis Patients and Centers* and *Special Information for Pet Stores and Aquarium Owners.*)

As this change will affect many people in the Tampa Bay region, some background on Tampa Bay Water and its operations may be helpful.

"Governance Agreements" Create Tampa Bay Water

In 1998, Tampa Bay Water became a wholesale water supply utility created to develop, store and supply water for Hillsborough, Pasco and Pinellas counties as well as the cities of New Port Richey, St. Petersburg and Tampa. They, in turn, distribute the water they purchase from Tampa Bay Water to their individual customers. Tampa Bay Water is a wholesale governmental supplier. It is funded through the sale of water and has no taxing authority.

In addition to being the region's wholesale water supplier, Tampa Bay Water was also charged with reducing adverse environmental effects caused from water withdrawals in concentrated areas. To accomplish this, a long-term Master Water Plan was developed which calls for diverse new water sources such as surface water, desalinated seawater and desalinated brackish water. A proposed regional reservoir will store surface water when there is an excess. Then, during drier times, this water can be used. In this way, Tampa Bay Water can significantly reduce groundwater pumping at the 11 regional wellfields and still meet the region's drinking water needs.

System Ownership Model

During the governance negotiations, a model of system ownership was agreed on whereby each member government continued to own and operate its existing drinking water treatment plants and distribution systems, while Tampa Bay Water purchased essentially all groundwater supply facilities and wellfield collection and transmission mains. Tampa Bay Water also became responsible for construction of any new water supplies. In other words, the individual city and county members of Tampa Bay Water own and operate all retail water distribution systems. Tampa Bay Water cannot own or operate any retail distribution system. It can only provide water on a wholesale basis to its member governments for distribution.

How the Partnership Works

To better understand this, imagine a water meter (see Figure 1 on next page). Tampa Bay Water is responsible for collecting, blending and treating water from all supply sources. As a wholesale water supplier, Tampa Bay Water delivers water treated to meet all Safe Drinking Water Act requirements, as well as additional parameters set forth in its own Governance Agreement, to its member governments (left side of the meter in Figure 1). Upon receiving that water, each member government then becomes responsible for final treatment and delivery to its customers (right side of meter in Figure 1).

Figure 1.

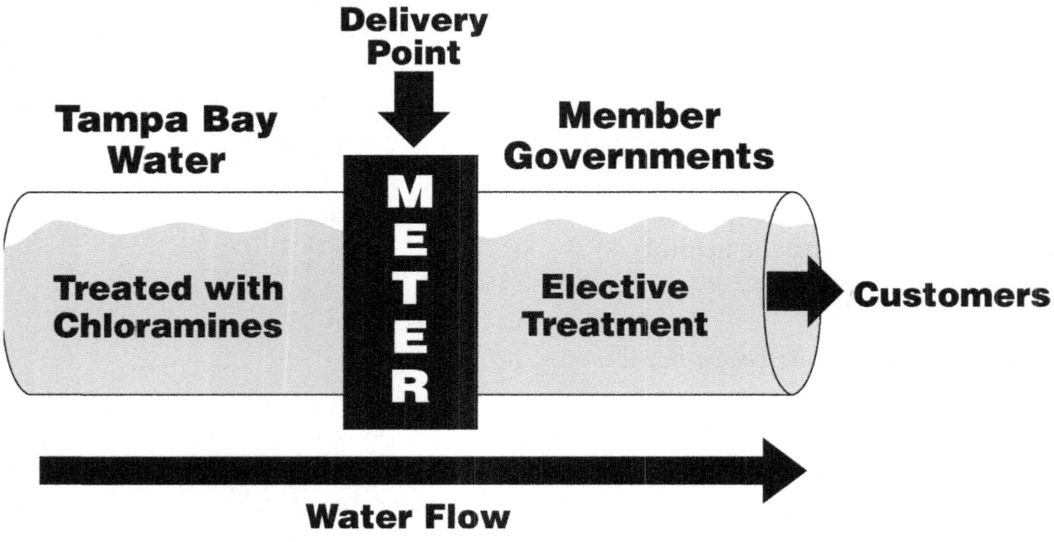

Each of the six member governments owns and operates its own water distribution system. This makes sense as each system is unique.

- Each was built at a different time
- Each is of a different size and layout
- Each has pipes made of different materials
- Each system's customer expectations based on individual experience are different

Because of these differences, each member government deals with corrosion control methods such as pipe replacement and water main flushing programs in its own way depending on when and how its system was built.

Member governments also retain ownership and operation of their own systems in order to ensure fair and equitable distribution of cost. If the responsibility for each distribution system were placed on Tampa Bay Water, the cost for operating and maintaining each system would be added to the wholesale water rate and charged to all members. But because each system's needs are different, adding this cost to the wholesale water rate would have unequal benefits for each member. By operating and maintaining its own distribution system, each member pays the same wholesale water rate and each member is responsible for final water disinfection, including the cost of the disinfection, as well as system maintenance.

Quality Water

In 1974, the United States Congress passed the Safe Drinking Water Act to protect public health by regulating the nation's public drinking water supply. The law was amended in 1986 and again in 1996. The 1996 amendments greatly enhanced the existing law, ensuring the quality of drinking water by protecting it from source to tap.

The Safe Drinking Water Act authorizes the United States Environmental Protection Agency (EPA) to set national health-based standards for drinking water to protect against both naturally occurring and man-made contaminants that may be found in drinking water. Every year research uncovers new microorganisms, and new technological procedures and tests aid the medical profession in uncovering previously unknown causes of disease. Based on this research in the water field, the EPA issues new water standard rules.

It is Tampa Bay Water's responsibility as a wholesale water provider to ensure that the water it delivers meets or is better than not only the regulations as defined by the EPA and the Safe Drinking Water Act, but also the water quality parameters required in its own Governance Agreements. Since 1998, several initiatives have been undertaken to scientifically ensure compliance with all these regulations as well as to provide guidance to both Tampa Bay Water for operation of its wolesale system, and to the member governments for operation of their retail distribution systems.

These initiatives include:

- Water sampling studies
- Studies for 10 additional water quality parameters
- Corrosion studies
- Establishment of water quality work groups
- A series of workshops on water quality issues presented to Tampa Bay Water's Board
- An approximately $3 million research project to study distribution system response to alternate source waters
- A wholesale blending analysis
- Design and implementation of chloramination systems in the wholesale system

The Decision to Use Chloramines vs. Free Chlorine

The wholesale water currently supplied by Tampa Bay Water is comprised of 99 percent groundwater and 1 percent surface water. By the end of 2007 the mix will have significantly changed. The area's water supply will be comprised of groundwater, surface water and desalinated seawater, each treated separately and then blended before delivery to the six member governments.

Both groundwater and surface water contain naturally occurring compounds referred to as natural organic matter. Decaying vegetation, for example, is a source of natural organic matter. Water must be treated to inactivate any disease-causing organisms that might make people sick. Disinfection occurs in two stages. The primary stage includes inactivation of any disease-causing organisms in the water at the water treatment plants. In the secondary stage, a small amount of disinfectant is maintained in the pipelines to inactivate any organisms that may recover from primary treatment and to protect against any contaminants that may enter the distribution system through main breaks, leaks or backflow. Both chlorine and chloramine are used as disinfectant residuals.

When water containing natural organic matter is treated with chlorine as its primary disinfectant, disinfection by-products occur. These by-products are regulated by the Safe Drinking Water Act. Scientists at the USEPA believe two primary types of regulated chlorination disinfection by-products — trihalomethanes (THMs) and haloacetic acids (HAAs) — may cause cancer with prolonged exposure.

In order to ensure ongoing compliance with more stringent health standards, the Tampa Bay Water Board of Directors in February 2000 decided to change its method of treating water. After looking at the advantages and disadvantages of six alternate combinations of disinfectant as well as different types of treatment processes, the decision was made to use chloramines rather than free chlorine to treat the water. By doing so, Tampa Bay Water joins dozens of other Florida utilities that use chloramines to protect against disease-causing organisms. Additionally, the use of chloramines reduces the level of THMs and HAAs found in water.

Member Governments Prepare to Receive Chloraminated Water

Before the conversion to chloramines by Tampa Bay Water, each of its member governments is taking steps to prepare for the change. These steps include:

1. Selecting a final distribution system disinfectant.
2. Implementing chemical feed system modifications.
3. Evaluating the impact of the new chemicals on its corrosion control and system flushing practices.
4. Engaging in re-permitting of its points of connection.
5. Participating in the development of a uniform public information approach.

The Team

David Carlson, PE
Black & Veatch Corporation

Mr. Carlson has 19 years of experience working primarily in the water treatment field, including numerous water supply, treatment and distribution projects that have involved process design and chemical feed systems. These projects have included the design of new facilities as well as rehabilitation of existing facilities. The treatment plants have ranged in capacity from 3 million gallons per day (mgd) to 75 mgd, using treatment processes such as clarification, softening, ion exchange and reverse osmosis. He has a bachelor's degree from Iowa State University and a master's degree from the University of Kansas.

Bruce Long, PE
Black & Veatch Corporation

Mr. Long has more than 30 years of water utility engineering experience with a bachelor's degree in chemical engineering from Lehigh University, a master's degree in environmental sciences from Rutgers University and is a Ph.D. candidate in environmental health engineering from the University of Kansas. He is responsible for assessing and developing design parameters for water treatment technologies for the firm's water treatment plant design projects. He also designs pilot plant testing facilities and test programs, evaluates the resulting data for upscaling to full-scale treatment processes, and provides assessments of existing treatment facilities as well as Safe Drinking Water Act compliance assessments.

Alan Hess, PE
Black & Veatch Corporation

Mr. Hess has more than 33 years of progressively responsible experience in water utility engineering, including management, design, studies, operation and maintenance at over 85 facilities. He has a bachelor's degree in civil and sanitary engineering from Pennsylvania State University and a master's degree in civil engineering from the University of Pennsylvania. His major areas of experience include: operation and maintenance of surface water and groundwater supply and treatment systems; compliance implementation with federal, state and local laws and regulations; organization, staffing, process design and optimization of new and existing water treatment facilities; research, development and implementation of innovative water treatment processes; water utility applications of instrumentation, control systems and computer technology. He has over 120 presentations and publications.

Tim Brodeur
Malcolm Pirnie, Incorporated

Mr. Brodeur has 29 years of experience in water and wastewater treatment process design, operations and trouble-shooting. He has a bachelor's degree and a master's degree in environmental engineering from the University of Florida. He has worked extensively with Florida utilities and other large systems throughout the United States in improving plant processes to comply with regulations including disinfection and control of disinfection by-products. He is currently managing a three-year contract to provide engineering support to the EPA on Safe Drinking Water Act regulations and how utilities will modify treatment plant processes to comply with new regulations and the cost of impact on the same.

In the course of his work with utilities and their change in disinfection practices, he has worked with over 15 systems, 10 of which involved changing from free chlorine to chloramines. Locally this included the city of Tampa in 1982. He is a Certified Water Treatment plant operator.

2535 Landmark Drive, Suite 211, Clearwater, FL 33761-3930
Phone: 727.796.2355 / Fax: 727.791.2388 / SunCom: 513.7010
www.tampabaywater.org

TAMPA BAY WATER
Supplying Water To The Region

Tampa Bay Water Quality Project

Chloramines and Free Chlorine

Two Methods of Disinfecting Water for Drinking Purposes

Water utilities throughout the United States have used both chlorine and chloramines for years as a secondary disinfectant for maintaining a disinfectant residual in distribution systems. Both methods, along with various other methods are effective in protecting public health and safety — the primary concern for any water supplier. And both methods are recognized throughout the United States as safe, effective means of disinfecting water. The choice as to which method to use basically comes down to which is more cost-effective for a particular utility under its unique circumstances.

The use of chlorine is an established technology. And, while chlorine is effective in controlling biological growth in a transmission system, chloramine is a more stable disinfectant, more persistent and stays in the distribution system longer, extending its disinfectant benefits. Both chlorine and chloramine are effective in controlling biofilm, a slime coating caused by bacteria that can occur in pipes.

While both methods are safe and effective, chloramines provide the following benefits:

- They are not as reactive as chlorine with organic material in water, resulting in substantially lower concentrations of regulated disinfection by-products, particularly those suspected to cause cancer such as trihalomethanes (THMs) and haloacetic acids (HAAs).
- Because the chloramine residual is more stable and lasts longer than chlorine, it provides better protection against bacterial re-growth in systems when using an appropriate flushing program.
- Because there will be less of a chlorine taste, water may taste and smell better.

2535 Landmark Drive, Suite 211, Clearwater, FL 33761-3930
Phone: 727.796.2355 / Fax: 727.791.2388 / SunCom: 513.7010
www.tampabaywater.org

TAMPA BAY WATER
Supplying Water To The Region

Tampa Bay Water Quality Project

Special Information for Kidney Dialysis Patients and Centers

In 2002, Tampa Bay Water will begin using chloramines rather than chlorine as its disinfectant residual in drinking water. This change is being made in advance of new federal and state drinking water regulations for disinfection by-products.

Chloramines are already being used throughout the country and are recognized as a proven, safe disinfectant residual. But like chlorine, chloramines must be removed from water to be used in kidney dialysis machines.

What does this mean for kidney dialysis patients?
Kidney dialysis patients can safely *drink, cook and bathe in* chloraminated water. Chloramines are harmful only when they directly enter the bloodstream through the dialysis process, and kidney dialysis patients may need to change the way water used for dialysis is pre-treated. Dialysis patients more than likely already test for both chlorine and chloramines, but depending on the method of chlorine removal a patient uses now, some modifications may be necessary. Patients should check with their dialysis physician who can recommend the proper type of water treatment.

Home users of kidney dialysis machines need to:

- Remove chloramines from water used for kidney dialysis treatment.
- Contact their dialysis center and complete any necessary modifications to their water treatment process. Often, home dialysis service companies can make the needed modifications, but patients should check with their physician to be sure.
- Check with their dialysis center to find out how often the center wants to see them. Monitoring for chloramine compounds in the water that has been pre-treated for dialysis may need to be performed more frequently.

Why do kidney dialysis patients need to take special precautions?
In the dialysis process, the compounds in water come in contact with the blood across a permeable membrane. Chloramines in that water would be harmful, just as chlorine is harmful, and must be removed from water used in kidney dialysis machines. This can be done by either adding ascorbic acid or by using a granular-activated carbon treatment. Medical centers that perform dialysis are responsible for preparing the water that enters the dialysis machines. They will be informed of this change in ample time to make any necessary adjustments.

Is it safe for kidney dialysis patients to drink water containing chloramines?
Yes. Because the digestive process metabolizes chloramines before they reach the bloodstream, everyone can drink chloraminated water. Kidney dialysis patients can *drink, cook and bathe* in chloraminated water. It is only when water interacts directly with the bloodstream, as in dialysis or in a fish's gill structure, that chloramines must be removed.

Can people with kidney ailments, with diabetes or on low-sodium diets use chloraminated water?
Yes. People with those medical problems can use chloraminated water for all everyday purposes such as bathing, drinking and cooking.

Will medical centers, hospitals and clinics that perform kidney dialysis know about the change to chloramines?
Yes. Tampa Bay Water will notify all medical facilities about the change from chlorine to chloramine disinfection. If people have any concerns about this process, they should talk with their physician.

July 25 2001

“Name”
“Title”
“Facility Name”
“Address 1”, “Address 2”
“City”, “ST” “Zip”

Subject: Alert Regarding Change in Water Treatment

Dear “Name”:

Please be alerted that two water treatment changes will be made in the near future that will require you to take appropriate measures to protect kidney dialysis patients.

1. Within the next month Hillsborough County is expected to begin providing water treated with **chloramines** to residents in northern Hillsborough County. This change will affect about 45,000 households. This change is being made to allow the county to use excess surface water from the City of Tampa when it is available, reducing its reliance on stressed groundwater sources.

2. In May, 2002 Tampa Bay Water, the region’s wholesale water supplier, will begin providing chloraminated water to its member governments, which in turn supply water to virtually every household in Hillsborough, Pinellas and Pasco counties. This change is being made in order to comply with stricter standards that the United States Environmental Protection Agency is expected to issue under the Safe Drinking Water Act.

Chloramines, a mix of chlorine and ammonia, are of particular concern to dialysis patients because they can cause hemolytic anemia, which can result in death. This potential danger can be avoided by treating the tap water with ascorbic acid or by using a granular-activated carbon filter to eliminate the chloramines.

It is essential that appropriate actions be taken to enable your facility, and any home-based patients that you may serve, to test for and remove chloramines from water before it is used for dialysis fluids.

Some local utilities, such as the City of Tampa, already use chloramines in their final disinfection process, so chloraminated water will not be new to everyone. In fact, your facility may already be testing for and/or removing chloramines on a regular basis.

Thank you for your attention to this important matter. If you have any questions please call .....

Sincerely,

Jonathan Kennedy, P.E.
Director of Facilities and Operations
Tampa Bay Water

2535 Landmark Drive, Suite 211, Clearwater, FL 33761-3930
Phone: 727.796.2355 / Fax: 727.791.2388 / SunCom: 513.7010
www.tampabaywater.org

TAMPA BAY WATER
Supplying Water To The Region

Tampa Bay Water Quality Project

Project Overview

In anticipation of new federal drinking water regulations, Tampa Bay Water will change its treatment process in 2002.

- Due to new research findings, the EPA will impose more strict drinking water regulations to protect consumers from suspected cancer-causing compounds.

- In order to comply, Tampa Bay Water will begin using chloramines as its secondary disinfectant prior to the regulations taking effect.

- The process has been used for decades across the country in places like Miami-Dade, Denver and Minneapolis. It is safe and the least expensive way to maintain compliance.

- Tampa Bay Water supplies water that meets regulations to its member governments, who, in turn, distribute water directly to consumers.

- Member governments must maintain compliance after they receive the water through a final step in the treatment process. If this final step differs from previous practices, consumers may notice a change in their drinking water.

- The change will consist of chlorine becoming much more difficult to detect. People most likely to notice the change will be residents of St. Petersburg, Pinellas and Pasco County and north Hillsborough County.

2535 Landmark Drive, Suite 211, Clearwater, FL 33761-3930
Phone: 727.796.2355 / Fax: 727.791.2388 / SunCom: 513.7010
www.tampabaywater.org

TAMPA BAY WATER
Supplying Water To The Region

Tampa Bay Water Quality Project

Questions & Answers

Why is Tampa Bay Water changing the way it treats drinking water?
While Tampa Bay Water already complies with all current water quality regulations, this change is being made in advance of new, stricter health standards expected to be imposed in the future under the national Safe Drinking Water Act. Chloramines were selected as the method of choice because they are the most cost-effective way of maintaining compliance under the new regulations. The use of chloramines is 20 percent less expensive than alternative methods of treatment, and at the same time reduces the formation of suspected cancer-causing compounds. And, by switching to chloramines, a more consistent water quality will be maintained throughout Tampa Bay Water's wholesale system.

The change from chlorine disinfectant to chloramine disinfectant will —

- Improve the overall disinfectant residual in Tampa Bay Water's wholesale system.
- Reduce the level of disinfection by-products formed when chlorine mixes with trace quantities of naturally occurring organic substances found in water, some of which are suspected to cause cancer with prolonged exposure.
- Provide the most economical treatment option available to Tampa Bay Water that will meet the new Safe Drinking Water Act standards.
- Minimize the chlorine taste and smell of drinking water.

Will all area cities and counties be affected by the change?
This change will not impact all Tampa Bay area residents in the same way. Any noticeable changes will depend on the type of final treatment each member government chooses to use and whether the chosen treatment differs from that member government's existing practices. A final treatment is necessary because the residual decreases with time and may disappear completely. For example, the city of Tampa already adds chloramines to its water as a final disinfectant, so its customers will not notice a change. Hillsborough County will begin providing chloraminated water to residents of north Hillsborough County and will continue providing water treated with free chlorine to residents of south-central Hillsborough County. Pinellas County, Pasco County and the city of St. Petersburg will begin treating their water with chloramines, so their residents may notice a change. New Port Richey, along with certain other areas that receive water from isolated sources, will not receive any chloraminated water from Tampa Bay Water, so those residents will not be affected. Municipalities that purchase water from Tampa Bay Water's members will be affected accordingly.

What steps do Tampa Bay Water's member governments need to take in order to accept chloraminated water?
Before the conversion to chloramines by Tampa Bay Water, each of its member governments is taking steps to prepare for the change. These steps include:

6. Selecting a final distribution system disinfectant.
7. Implementing chemical feed system modifications.
8. Evaluating the impact of the new chemicals on its corrosion control and system flushing practices.
9. Engaging in re-permitting of its points of connection.
10. Participating in the development of a uniform public information approach.

Is anything else added to the water before delivery to customers?
Member governments may elect to use different additional treatments such as fluoride, softening agents, pH adjustments, etc. While Tampa Bay Water doesn't add any of these to the water it supplies to its member governments, Hillsborough County, St. Petersburg and Tampa add fluoride to the water before delivery to customers, and both St. Petersburg and Tampa soften their water.

Will I notice a difference in the taste or odor of my water?
Residents who will be receiving chloraminated water for the first time may notice a change in taste or odor. Chlorine will become much more difficult to detect.

Will there be any noticeable difference in my water due to the switch to chloramines?
Some temporary impacts may be noticed in isolated areas as the chloraminated water is introduced into the member governments' retail distribution systems. Some customers may notice a temporary variation in water color as well as possible sediment in the water. These temporary impacts to water quality are similar to those experienced when a water main is replaced or routine maintenance is performed on the water distribution system and will cease as the system stabilizes.

I have a fish aquarium in my home. Am I going to have to do anything different to the water?
Yes. Since fish take chloramines directly from the water into their bloodstream, the chloramines must be removed from the water used by your fish and other aquatic animals. This process is similar to what you may already be doing to remove chlorine from your aquarium water. Some people, however, may simply let the water sit out before use so the chlorine will dissipate. *Chloramines, however, will not dissipate in this manner.* Water conditioners specifically designed for removing both chlorine and chloramines are commercially available. Pet stores or pet suppliers should be able to provide information on de-chloramination products and instructions. (A more thorough answer to this question can be found in the *Special Information for Pet Stores and Fish Aquarium Owners* insert.)

Will the use of chloramines affect kidney dialysis centers and patients?
Yes. Kidney dialysis patients must remove chloramines from the water they use in dialysis machines. Depending on the method of chlorine removal a patient now uses, some modifications to this process may be necessary to remove chloramines. Patients should check with their dialysis physician who can recommend the proper type of water treatment. Dialysis centers will be notified in ample time to make any necessary changes to their equipment. (A more thorough answer to this question can be found in the *Special Information for Kidney Dialysis Patients and Centers* insert.)

Is this change going to cost me more on my water bill?
This change will have a nominal impact (a 2-3 cent increase per thousand gallons) on the wholesale cost to produce the water. This will affect the rate your local utility is charged. Whether this change will be reflected on *your* water bill will be up to each member government to decide.

Will chloraminated water be safe for my dog or cat?
Yes.

When will the change be implemented?
May 2002.

2535 Landmark Drive, Suite 211, Clearwater, FL 33761-3930
Phone: 727.796.2355 / Fax: 727.791.2388 / SunCom: 513.7010
www.tampabaywater.org

TAMPA BAY WATER
Supplying Water To The Region

Tampa Bay Water Quality Project

Glossary

Carcinogenity — A term referring to the likelihood of a substance causing cancer.

Chloramines — A mix of chlorine and ammonia in water that is used to disinfect water for drinking water purposes.

Corrosion — Refers, in this case, to possible wearing away through chemical action of the inside of water distribution pipelines due to age and chemical make-up of water flowing through the pipes.

Disinfection — The inactivation of disease-causing organisms in water. Disinfection occurs in two stages. The primary stage includes inactivation at the water treatment plants. In the secondary stage a small amount of disinfection is maintained in the pipelines to inactivate any organisms that may recover from primary treatment or be introduced into the distribution system through main breaks, leaks or backflow.

Disinfection By-Product — When water containing natural organic matter is treated with any disinfectant, disinfection by-products that are regulated by the Safe Drinking Water Act occur. The Safe Drinking Water Act authorizes the United States Environmental Protection Agency (USEPA) to set national health-based standards, for drinking water to protect against both naturally occurring and man-made contaminants that may be found in drinking water. These regulated disinfection by-products include trihalomethanes (THMs) and haloacetic acids (HAAs).

Disinfection Residual — This term refers to the practice of maintaining small amounts of a disinfectant in the water at all times to inactivate organisms that can make people sick. The residual may be either free chlorine or chloramines.

Free Chlorine — Refers to the use of chlorine only to disinfect water for drinking purposes.

Governance — This is the term that refers to the agreements that established Tampa Bay Water as a wholesale water supply utility. In this agreement, Tampa Bay Water became the sole owner of essentially all groundwater supply facilities and wellfield collection and transmission mains of its member governments as well as being responsible for construction of new water supplies.

Natural Organic Matter — Natural organic matter refers to naturally occurring compounds in both groundwater and surface water. Decaying vegetation, for example, is a source of natural organic matter.

2535 Landmark Drive, Suite 211, Clearwater, FL 33761-3930
Phone: 727.796.2355 / Fax: 727.791.2388 / SunCom: 513.7010
www.tampabaywater.org

TAMPA BAY WATER

Tampa Bay Water Quality Project

Special Information for Pet Stores and Fish Aquarium Owners

In 2002, Tampa Bay Water will begin using chloramines rather than chlorine as the disinfectant residual in its drinking water. This change is being made in advance of new federal and state drinking water regulations for disinfection by-products. Chloramines have been successfully used throughout the country for decades.

What does this mean for pet stores, fish aquarium owners and pond owners?
Chloramines are only harmful when they directly enter the bloodstream. Since fish and other aquatic animals take chloramines directly from the water into their bloodstream, the chloramines must be removed from the water used for your aquarium. This process is similar to what you may already do to remove chlorine. Leaving the water out in an open container, however, will not dissipate chloramines like it may dissipate chlorine. Water conditioners specifically designed for removing chloramines are commercially available. Pet stores or pet suppliers should be able to provide you with de-chloramination products and instructions.

How do chloramines affect fish?
Chloramines are harmful to fish and must be removed from water just as chlorine is harmful and must be removed. While you might not have to remove chlorine from your aquarium water because it breaks up rapidly on its own, this is not the case with chloramines. Most pet stores have sold de-chlorinating agents for years and have generally recommended their use. These chemicals used to remove chlorine should work just as well for chloramines. Several manufacturers have had chloramine information on their product labels for years.

How does ammonia affect fish?
When chloramines are chemically removed, ammonia is also released. Ammonia can be harmful to fish even though fish produce some ammonia as a by-product. Although ammonia levels may be tolerable in individual tanks or ponds, commercial products are available to remove excess ammonia. Biological filters, natural zeolites and pH control methods are also useful in reducing the toxic effects of ammonia.

Are both saltwater fish and freshwater fish affected by chloramines?
Yes. Chloramines affect saltwater fish just as they affect freshwater fish. Chloramines must be removed if the water used to make saltwater solutions comes from a chloraminated supply.

Are Koi fish affected by chloramines like other fish?
Yes. Koi are just as susceptible to being harmed by chloramines as any other fish.

Won't letting water sit for a few days remove chloramines from tanks or pond water?
No. Unlike chlorine, which dissipates when water sits for a few days, chloramines are longer lasting and may take weeks to dissipate. If people don't want to use a de-chloraminating chemical, the next best solution is to install a granular-activated carbon filter and allow sufficient contact time between the water and filter. Follow the filter's instructions.

How much of a de-chloraminated agent or what type of granular-activated carbon filter should be used?
Check the product label for specific instructions or ask a pet supplier.

Attention Fish Owners!

Your Tap Water Is Changing.

Here's How You Can Protect Your Fish:

Beginning October 1, 2001, water in certain areas of Northwest Hillsborough County will be treated with chloramines instead of chlorine. Chloramines do not dissipate as readily as chlorine. They must be removed from your tap water with a water conditioning agent or filter before using it in your aquarium or fishpond. If not neutralized, your fish could die.

Products designed to remove chloramines are available. Ask your pet supply store or your local utility for more information.

What Fish Owners Need to Know about Chloramines

Much of the water supplied to Hillsborough, Pasco and Pinellas counties from Tampa Bay Water, the region's wholesale water supplier, will soon be treated with chloramines. This will help Tampa Bay Water, the regional water supply wholesaler, and your local water utility meet new, stricter water regulations set forth by the Environmental Protection Agency. Currently, 20-25 percent of local governments in the United States use chloramines, and the number is expected to grow due to new federal water quality regulations.

What does this mean for aquarium and pond owners?
While the chloraminated water that customers receive will be superior to state and federal drinking water standards, chloramines are harmful to fish when they directly enter the bloodstream. Since fish and other aquatic animals take chloramines directly from the water into their bloodstreams through their gills, **chloramines must be removed from water used for keeping live fish and other aquatic life.**

What types of aquatic life do chloramines affect?
Chloramines, like chlorine, will kill both salt and fresh water fish and other aquatic life including Koi fish, lobster, shrimp, frogs, turtles, snails, clams and live coral. Chloramines, therefore, like chlorine, must be removed before using tap water in your fresh or salt-water aquarium or pond.

How can I remove chloramines from my water?
A water-conditioning agent specifically designed to remove chloramines or an activated carbon filter must be used according to product instructions. If you are already using one of these products to remove chlorine, it's possible that the same product will also remove chloramines. However, **you must read the product label** to be sure. Chlorine removal agents that are not specifically designed to also remove chloramines could leave excess ammonia in the water. Too much ammonia could harm fish. Pet stores and other retail outlets have sold these products for years and have generally recommended their use. Your pet supplier should be able to provide any further guidance you may need.

Won't letting water sit for a few days remove chloramines from tanks or pond water?
No!
Unlike chlorine, which dissipates when water sits for a few days, chloramines are longer lasting and may take weeks to dissipate. This is not a safe method for removing chloramines.

For more information, go to www.tampabaywater.org or call the Hillsborough County Lake Park water plant any time at 813-264-3835. Operations staff is available 24 hours a day, seven days a week.

TAMPA BAY WATER
Supplying Water To The Region

OVERVIEW
Public Information Program

Communications Vehicles

- Fact Pack
  - Inform
  - Gather input/Put it in writing
  - Provide common reference material
  - Reduce speculation and misinformation
  - Facilitate accurate communications
  - Modular/Dynamic/Flexible

Communications Vehicles

- Web site
- Consumer brochure/Utility bill stuffer
- Sample newsletter story
- Speaker's bureau

Communications Vehicles

- Letters to dialysis centers/fish suppliers
- Pet store brochure/signage
- Multi-media ad campaign
  - starts in March
- Call Center
  - starts in March

Fact Pack Contents

- Project overview
- Background
- Chloramines and free chlorine
- White paper
- Information targeted to two special groups
- Questions & answers
- Glossary

QUESTIONS

IWA Publishing's authorised EU representative for General Product Safety Regulations is Diane D'Arras, 15 rue Duret, 75116 Paris, France, e-mail: safety@iwap.co.uk.

Printed and bound by CPI Group (UK) Ltd, Croydon, CR0 4YY
17/09/2026
02218811-0020